Lecture Notes in Mathematics 1560

Editors:
A. Dold, Heidelberg
B. Eckmann, Zürich
F. Takens, Groningen

Thomas Bartsch

Topological Methods for Variational Problems with Symmetries

Springer-Verlag

Berlin Heidelberg New York
London Paris Tokyo
Hong Kong Barcelona
Budapest

Author

Thomas Bartsch
Mathematisches Institut
Universität Heidelberg
Im Neuenheimer Feld 288
D-69120 Heidelberg, Germany

Mathematics Subject Classification (1991): 34Cxx, 35J20, 47H15, 49R05, 55M30, 55P91, 58Exx, 58Fxx

ISBN 3-540-57378-X Springer-Verlag Berlin Heidelberg New York
ISBN 0-387-57378-X Springer-Verlag New York Berlin Heidelberg

2146/3140-543210 - Printed on acid-free paper

meinen Eltern

Preface

Symmetry has a strong impact on the number and shape of solutions to variational problems. This can be observed, for instance, when one looks for periodic solutions of autonomous differential equations and exploits the invariance under time shifts or when one is interested in elliptic equations on symmetric domains and wants to find special solutions.

Two topological methods have been devised in order to find critical points that are not minima or maxima of variational integrals: the theories of Lusternik-Schnirelmann and of Morse. In these notes we want to present recent techniques and results in critical point theory for functionals invariant under a symmetry group. We develop Lusternik-Schnirelmann (minimax) theory and a generalization of Morse theory, the Morse-Conley theory, in some detail. Both theories are based on topological notions: the Lusternik-Schnirelmann category and geometrical index theories like the genus on the one hand and the Conley index (generalizing the Morse index) on the other hand. These notions belong to the realm of homotopy theory with a typical consequence: They are easy to define but difficult to compute.

We present a variety of new computations of the category where very general classes of symmetry groups are involved, and we give examples showing that our results cannot be extended further without serious restrictions. In order to do this we prove new generalizations of the Borsuk-Ulam theorem and give counterexamples to more general versions. It is here that we need to use some algebraic topology, namely cohomology theory, and an equivariant version of the cup-length.

A variation of the equivariant cup-length, the "length", turns out to be very useful also in our presentation of the Morse-Conley theory for symmetric flows. The length is a cohomological index theory. It should be considered as a measure for the size of invariant sets. We use it to associate an integer, the "exit-length", to an isolated invariant subset of a flow. The exit-length is closely related to the Morse index of a nondegenerate critical point. We apply our general theory to bifurcation problems and show that a change of the exit-length along a branch of stationary solutions gives rise to bifurcation. The length of the bifurcating set of bounded solutions can be estimated, and this in turn can be used to analyze the bifurcating set. This approach to bifurcation theory will be illustrated with two applications, namely the bifurcation of steady states and heteroclinic orbits of $O(3)$-symmetric flows, and with the existence of periodic solutions near equilibria of symmetric Hamiltonian systems. We discuss the symmetry of the bifurcating solutions, obtain multiplicity results and discover special solutions. For example, we find brake orbits and normal mode solutions for certain symmetric Hamiltonian systems.

I wish to thank all those who have furthered my understanding of the subject and helped me during the preparation of these notes. In particular I want to mention Mónica Clapp, Albrecht Dold, Ed Fadell, Bernold Fiedler, Hans-Werner Henn, Reiner Lauterbach and Dieter Puppe.

Contents

Chapter 6. The length and Conley index theory

Chapter 7. The exit-length

Chapter 8. Bifurcation for O(3)-equivariant problems

Chapter 9. Multiple periodic solutions near equilibria of symmetric Hamiltonian systems

Chapter 1

Introduction

1.1 The question

The "least action principle" plays an impressive role in the development of modern science. Its origin dates back to the work of the Greek on isoperimetric problems. Formulated explicitly by Euler and Maupertuis in 1744 it led to the foundation of a new mathematical discipline, the Calculus of Variations. Since the times of Euler many mathematicians took part in the development of this discipline: Lagrange, Legendre, Jacobi, Weierstraß, Hilbert, Lebesgue, to mention a few. Looking at recent results on minimal surfaces or metrics with prescribed curvature properties, harmonic maps of Riemannian manifolds, Hamiltonian systems or symplectic geometry, it seems that the Calculus of Variations is as vital today as ever.

Compared with this history the search for critical points other than extrema of variational integrals is relatively young. It begins with the famous work of Birkhoff in 1917 and of Lusternik, Schnirelmann and Morse in the 1920s on the existence of closed geodesics on compact Riemannian surfaces of genus zero. Their work is based on the observation that the total manifold of functions on which the variational integral is defined can be decomposed into pieces which depend on the number and type of the critical points and on the gradient flow of the functional. In the approach of Lusternik and Schnirelmann the pieces are contractible subsets which leads to the so-called Lusternik-Schnirelmann category. In the approach of Morse the pieces are cells and the homology of this cell decomposition gives the homology of the manifold. Both approaches have been extended and applied to numerous problems from mathematical physics to differential geometry and topology. In fact, recent versions of the Lusternik-Schnirelmann and Morse theory are now capable to deal with variational principles which — a few years ago — appeared to be too degenerate to be of any use. The work of Rabinowitz and Floer, in particular, on Hamiltonian systems and symplectic geometry is a model example for this development.

Already in the early days of minimax theory Lusternik and Schnirelmann observed that symmetry plays an important role. Around 1930 they proved that an even C^2-map $S^n \to \mathbb{R}$ must have at least $n+1$ pairs $\pm x$ of critical points. Whereas it is clear that critical points of even maps occur in antipodal pairs it is one of the most fascinating features of critical point theory that symmetry forces the existence of many more critical points in addition to the minimum and the maximum. The topological result needed in the proof is the computation of the Lusternik-Schnirelmann category $\mathrm{cat}(\mathbb{R}\mathrm{P}^n)$ of the orbit space $\mathbb{R}\mathrm{P}^n = S^n/(x \sim -x)$. This is the minimal cardinality of a covering of $\mathbb{R}\mathrm{P}^n$ by open subsets which can be deformed to a point inside $\mathbb{R}\mathrm{P}^n$. Indeed, $\mathrm{cat}(\mathbb{R}\mathrm{P}^n)$ is a lower bound for the number of critical pairs as indicated above. The equality $\mathrm{cat}(\mathbb{R}\mathrm{P}^n) = n+1$ is an easy consequence of the Borsuk-Ulam theorem which states that there does not exist an odd map $S^n \to S^{n-1}$.

Replacing the symmetry group $\mathbb{Z}/2 = \{\pm \mathrm{Id}\} \subset O(n+1)$ by more general groups, one is led to the following questions:

> Suppose the C^2-map $\Phi: S^n \to \mathbb{R}$ is invariant under the action of a subgroup G of $O(n+1)$, that is $\Phi(gx) = \Phi(x)$ for every $x \in S^n$ and $g \in G$. How many G-orbits of critical points does Φ have? In particular, does Φ have at least $n+1$ such critical G-orbits if G is finite and if the action is fixed point free?

Here a point $x \in S^n$ is said to be a fixed point of the action if $gx = x$ for every $g \in G$. It is clear that one has to exclude such fixed points because otherwise the orthogonal projection $S^n \subset \mathbb{R}^{n+1} \to \langle x \rangle \cong \mathbb{R}$ onto a subspace spanned by a fixed point provides an invariant map having only two critical points, namely its maximum x and its minimum $-x$. Again it is easy to see that the set of critical points is G-invariant, that is if $x \in S^n$ is a critical point of Φ then so is gx for every $g \in G$. But very little is known about the number of these critical G-orbits. A lower bound for this number is the category of the orbit space $S^n/G = S^n/\sim$, where $x \sim gx$ for $x \in S^n$ and $g \in G$. And related to the computation of $\mathrm{cat}(S^n/G)$ are generalizations of the Borsuk-Ulam theorem about the existence or rather non-existence of certain maps which respect the action of G.

Why should anyone be interested in these problems apart from the fact that they are "natural" generalizations of old and beautiful results? Starting with the theorem of Lusternik and Schnirelmann a critical point theory for symmetric functionals has been developed and applied very successfully to many problems from the calculus of variations. To mention a few results in this direction:

— infinite-dimensional problems with constraints, like the non-linear eigenvalue problem $\nabla \Phi(x) = \lambda x$ where $\Phi: X \to \mathbb{R}$ is a G-invariant functional defined on a Hilbert space X with a linear isometric action of G (here one has to find critical points of Φ constrained to the unit sphere of X);

— the bifurcation theory for non-linear symmetric potential operators, that is solutions of $\nabla \Phi_\lambda(x) = 0$ where $\Phi_\lambda: X \to \mathbb{R}$ is a family of functionals as described above satisfying $\nabla \Phi_\lambda(0) = 0$ for every $\lambda \in \mathbb{R}$;

— the symmetric mountain pass theorem and more general linking theorems;

— applications of these results to elliptic partial differential equations;

— applications to the existence of periodic solutions of Hamiltonian systems (on a given energy surface or with a given period or near a stationary solution).

Since the literature on this subject is vast and is still growing enormously we only mention three of the most recent books respectively surveys where one can find further references concerning all the topics (and more) mentioned above. These are Struwe's monograph [Str] on "Variational Methods" with applications to nonlinear partial differential equations and Hamiltonian systems; the book [MaW] by Mawhin and Willem on "Critical Point Theory and Hamiltonian Systems" which contains many historical and bibliographical notes; and the survey [Ste] of Steinlein on the topology behind the theorem of Lusternik and Schnirelmann, namely on "Borsuk's Antipodal Theorem and its Generalizations and Applications".

In most papers in this field the attention is restricted to the cyclic groups $\mathbb{Z}/p$ of prime order or to the group S^1 of complex numbers of modulus 1. And in almost all of those papers which deal with more general groups there is a restrictive assumption which allows to deduce the general result from the corresponding $\mathbb{Z}/p$-version or S^1-version. Typical assumptions of this type are that the action of G (on S^n or $X-0$) is free, or there exists a subgroup $H \cong \mathbb{Z}/p$ of G which acts freely, or there exists a prime p which divides the Euler characteristic of each orbit $Gx = \{gx: g \in G\}$ (for G finite this is just the number of elements of Gx). One can also find the definition of index theories for arbitrary groups. However they are only computed if $G = \mathbb{Z}/p$, if $G = S^1$ or if additional restrictions on the action as those described above are satisfied.

One of the motivations for this research was the desire to understand whether assumptions of this kind are necessary. For example, we address the problem of finding necessary and sufficient conditions on Banach spaces X with a linear isometric action of G so that $\mathrm{cat}(SX/G)$ is infinite. This allows us to extend the eigenvalue theory for odd nonlinear operators to a much larger class of symmetric operators. Studying this and related problems we shall prove Borsuk-Ulam type theorems which hold for $\mathbb{Z}/p$ but not for $\mathbb{Z}/(p^k)$, $k > 1$, which hold for p-groups but not for solvable groups, and which hold for solvable groups but not for non-solvable groups. In this way we get some feeling for the assumptions on the action needed in applications. It becomes clear that it is important to know precisely which version of the Borsuk-Ulam theorem one needs when one wants to generalize a result from $\mathbb{Z}/2$ to other symmetry groups. This will be demonstrated, for instance, with a version of the symmetric mountain pass theorem for a very general class of groups and an application to an elliptic boundary value problem.

There are plenty of "real life" applications where much more complicated symmetry groups than $\mathbb{Z}/p$ or S^1 appear. For instance, the spherical pendulum has an $O(2)$-symmetry, and if one translates the problem of finding periodic solutions into a variational setting then one has to deal with the group $S^1 \times O(2)$. The S^1-action comes from the time shifts on a space of periodic functions. Similarly the Hénon-Heiles Hamiltonian is invariant with respect to the dihedral group D_3 of 6 elements, which leads to the symmetry group $S^1 \times D_3$. And every natural Hamiltonian $H(p,q) = \frac{1}{2}\|p\|^2 + V(q)$ is invariant under the reflection $(p,q) \mapsto (-p,q)$. Since this action of $\mathbb{Z}/2$ is anti-symplectic it combines with the time shifts to an action of the semidirect product $S^1 \rtimes \mathbb{Z}/2 \cong O(2)$. Examples of a completely different kind are the buckling of a spherical shell or the convection of a fluid confined between two spherical shells which are modeled by $O(3)$-invariant partial differential equations. More generally, an elliptic equation defined on a domain Ω in $\mathbb{R}^n$ inherits the symmetry of the domain. In the two examples just mentioned this is the group $O(3)$, but other subgroups of $O(n)$ appear also very naturally.

Indeed, these problems have already been attacked using various methods, most notably methods from degree theory or singularity theory. In the PDE applications just mentioned one is not only interested in the steady states of the system (which correspond to the critical points of the potential) but also in heteroclinic solutions of the associated parabolic equation. More complicated dynamics like periodic solutions or homoclinic orbits cannot occur in systems of variational type. This suggests that one should really study gradient or gradient-like flows which are symmetric with respect to some symmetry group G. In order to do this we present a version of the Morse-Conley theory which works very well in particular for "nice" groups. These results are then applied to bifurcation problems where "bad" groups are involved, like the bifurcation of steady states and connecting (heteroclinic) orbits for $O(3)$-equivariant problems; and the existence of multiple periodic solutions of symmetric Hamiltonian systems near a stationary solution where we have to deal with symmetry groups of the form $S^1 \rtimes G$.

1.2 The approach

Our program consists of three steps:

1. We extend the geometric notion of Lusternik-Schnirelmann category to the equivariant world and develop some critical point theory for functionals with symmetries (in chapter 2).

2. We use algebraic topology to compute the equivariant category for spheres in representation spaces of G. To do this we define a cohomological index theory for spaces with group actions in analogy to the cup-length (in chapters 3-5).

3. We use this cohomological index in combination with the Conley index to study gradient-like flows with symmetries in situations where more geometric notions like the category do not seem to be applicable (in chapters 6-9).

The basic version of the equivariant Lusternik-Schnirelmann category of a G-space X, G-cat(X), is not the category of the orbit space X/G. If one wants the category to be maximal with respect to certain properties (deformation invariance, subadditivity, normalization), one is led to the following definition: G-cat(X) is the smallest cardinality of a covering of X by G-invariant open subsets, which can be deformed equivariantly to a G-orbit. A G-orbit is a homogeneous G-space and should be considered as an "equivariant point". This version of the category is due to Fadell [Fa2]. Various extras can be added which are useful in critical point theory. As in the non-symmetric case it is difficult to compute G-cat(X). We are mainly interested in the category of the unit sphere $X = SV$ of a (possibly infinite-dimensional) representation space of G. One of our results is a classification of those compact Lie groups which satisfy G-cat$(SV) = \infty$ for every infinite-dimensional representation V of G (provided there are no fixed points in SV; otherwise G-cat$(SV) \le 2$).

The following observation is useful for computations: If there exists an equivariant map $SV \to SW$, then G-cat$\big(S(V \oplus W)\big) \le 2 \cdot G$-cat$(SW)$. Thus we have to study the existence or non-existence of equivariant maps between representation spheres, which is the content of Borsuk-Ulam type theorems. For instance, if one can construct a G-map $SV \to SW$ with $\dim V = \infty$ and $\dim W < \infty$, then the infinite-dimensional representation sphere $S(V \oplus W)$ has finite G-category. This also implies that the non-equivariant category of the orbit space $S(V \oplus W)/G$ is finite. At first sight this is surprising because $S(V \oplus W)/G$ is something like an infinite-dimensional projective space. Results of this type (which we prove in chapter 3) are bad from the critical point theory point of view. On the other hand, in this way one can obtain new estimates for the category of complicated spaces like orbit spaces of representation spheres.

If $G = \mathbb{Z}/2$ then two other topological invariants have been introduced in the literature: the genus of Krasnoselski [Kr1,2] and Yang [Y2] (who called it B-index), and the cohomological index of Fadell and Rabinowitz [FaR1]. This has also been defined by Yang [Y1], but he did not apply it to critical point theory. These notions coincide for representation spheres but differ in general. One always has

$$(\mathbb{Z}/2)\text{-category} \geq \text{genus} \geq \text{cohomological index.}$$

The genus and the cohomological index are examples of so-called index theories. They are useful as lower bounds for the category. But they have also been applied directly to critical point theory because they have better properties than the category. In particular the cohomological index has been applied by Fadell and Rabinowitz very successfully to bifurcation theory, and it is doubtful whether one can prove their results using the genus or the category instead. The genus and the cohomological index have been extended to more general groups. Here we want to mention the work of Fadell, Husseini and Rabinowitz ([Fa1-3], [FaH1,2], [FaR2]) and a paper of Benci [Be2]. Applications and computations exist mainly for the group S^1.

We shall define such index theories for arbitrary groups. Our version of the genus coincides with the one of Krasnoselski and Yang for $G = \mathbb{Z}/2$ and with the geometrical index of Benci for $G = S^1$. Similarly, our cohomological index coincides with the one of Fadell and Rabinowitz for $\mathbb{Z}/2$ and S^1. For other groups it is different and yields better results. (Already for tori the index theory of Fadell and Rabinowitz is not really applicable and one has to use the ideal-valued index defined in [FaH2].) It is called the length because it is closely related to the cup-length. The length is a slight modification of the version introduced and applied by Clapp, Puppe and the author ([ClP2], [BC1], [BCP]). The length is used to obtain lower bounds for category and genus. We compute it in a number of cases and not only for $\mathbb{Z}/2$ or S^1. One of the reasons that we are able to obtain results for other groups is that we work not only with singular cohomology theory. For example, in order to obtain lower bounds for $G\text{-cat}(SV)$ with $G = \mathbb{Z}/(p^k)$ a cyclic p-group we apply equivariant K-theory. Furthermore if G is an arbitrary p-group we use equivariant stable cohomotopy. These computations are collected in chapter 5.

As with the cohomological index of Fadell and Rabinowitz the length has additional properties which do not hold for the category nor the genus. We shall need these properties in order to study gradient-like flows with symmetries. This will be done with the help of an equivariant version of the Morse-Conley theory. We shall show that the lengths of an isolated invariant set of a flow, of its Conley index and of its unstable set are interesting invariants. We concentrate on bifurcation theory where we are able to estimate the length of the bifurcating invariant set from below, thus obtaining a measure of its size. The chapters 6 and 7 are devoted to this topic. In order to illustrate the questions which we investigate, consider a family of flows $\varphi_\lambda = \varphi_\lambda(x,t)$ on $X = \mathbb{R}^n$. Suppose φ_λ is gradient-like and odd (with respect to $x \in X$). Then the origin is a stationary solution of φ_λ for every $\lambda \in \mathbb{R}$. Assume dimension of the unstable manifold of 0 (with respect to the flow φ_λ) changes from d_- to d_+ as λ passes λ_0. Then it is clear that λ_0 is a bifurcation point. For the sake of exposition let us also assume that all bifurcating "branches" of stationary solutions are supercritical. In that case Fadell and Rabinowitz [FaR1] proved the existence of $d = |d_+ - d_-|$ antipodal pairs $\pm x_1, \ldots, \pm x_d$ of stationary solutions of φ_λ for $\lambda > \lambda_0$ close to λ_0. (In fact they only considered flows associated to the ODE $\dot{x} = \lambda x - \nabla\Phi(x)$ with $\Phi: X \to \mathbb{R}$ even.) Using methods from Conley index theory combined with the length we shall prove the existence of a compact flow invariant (with respect to φ_λ) and $\mathbb{Z}/2$-invariant set $S_\lambda \subset X - 0$ whose length is at least d. This implies the result of Fadell and Rabinowitz on the number of stationary solutions. In addition S_λ must contain heteroclinic orbits connecting the bifurcating stationary solutions x_{k+1} and x_k, for $k = 1, \ldots, d-1$. We also obtain results on the critical groups of the stationary solutions and on the dimension of S_λ. Moreover, the set S_λ is stable under small perturbations in the sense that the length of its $\mathbb{Z}/2$-Conley index is at least d.

Similar results will be proved for $\mathbb{Z}/p$ or S^1, for instance, but the discussion of chapters 3 to 5 shows that one cannot expect such a theory for arbitrary compact Lie groups. The reason is that for $G = \mathbb{Z}/2$ the dimension of the unstable manifold of the hyperbolic stationary solution $0 \in X$ is the same as the length of its exit set. Indeed, the exit-set is G-homotopy equivalent to the sphere of the tangent space of the unstable manifold at 0, and in chapter 3 we showed that there is in general no relation between $\dim V$ and the length (or category or genus) of SV.

A way out of this dilemma will be illustrated by two case studies. In chapter 8 we use the $\mathbb{Z}/2$-equivariant bifurcation theory to deal with $O(3)$-symmetric problems. Here we assume that we have performed a center manifold reduction so that all bounded solutions of the original partial differential equation near a bifurcation point can be found by studying a flow on a finite-dimensional representation space of G. We investigate the bifurcation of steady states and heteroclinic orbits, their symmetries and multiplicities. In chapter 9 we apply our results for tori $(S^1)^k$ to find periodic solutions of symmetric Hamiltonian systems near a stationary solution. We prove generalizations of the Weinstein-Moser theorem [We], [Mos], on periodic solutions which lie on energy surfaces near the stationary solution. Moreover we obtain results on the number of periodic solutions with fixed period close to the periods of the linearized system, thus generalizing the work of Fadell and Rabinowitz [FaR1,2]. We also study the bifurcation of special solutions like brake orbits or normal mode solutions for certain symmetric Hamiltonian systems. Our presentation of these subjects is certainly not exhaustive, not even with respect to the possibilities of the method which we develop. We just want to illustrate the general theory, and we want to give food for future thoughts.

1.3 An advice

For those readers who have followed so far it is evident that we use a fair (?) amount of topology. We tried to make the results accessible to all those who are interested in variational problems, Conley index theory and applications, and who prefer to avoid the more algebraic parts of topology. Certainly one should feel comfortable with concepts like the category or the genus and their relation to the cuplength. And one should be acquainted with the basic expressions of the equivariant language as collected in the first paragraphs of §2.2. More complicated machinery is being used in chapters 3, 4 and 5. These parts can essentially be skipped. Indeed, one can pass from §3.2 where we state the results on the category and the genus of representation spheres directly to chapter 6 — with one exception: One should have a look at §4.1 and at §4.4 where we state (and prove) the most important properties of the length. There we only use the formal properties of cohomology theories (the Eilenberg-Steenrod axioms) and of the cup-product. Of course, it won't hurt to read §§4.2, 4.3 up to definition 4.1 of the length and the introduction to chapter 5, too, just to get an idea of what goes into the computation of the length. There are a few results in chapters 6 to 9 whose proofs need more topology. One can skip these proofs and (almost) replace the length by the genus.

On the other hand, if one is mainly interested in the computation of the Lusternik-Schnirelmann category or the genus and in the results about equivariant maps between representation spheres, then one can go directly to chapters 3 to 5 after having a glance at the definitions 2.6 and 2.8 and at §2.4.

Chapter 2

Category, genus and critical point theory with symmetries

2.1 Introduction

After an introduction to equivariant topology we define and study the general concept of Lusternik-Schnirelmann category of a map in the equivariant context which is due to Clapp and Puppe [ClP1,2]. In the classical version the Lusternik-Schnirelmann category of a space X asks for the minimal number of open subsets needed to cover X and such that each element of this covering can be deformed (in X) to a point (cf. [LuS2]). If a group G acts on X then one wants the covering and the deformations to respect the action. For instance, if $G = \mathbb{Z}/2$ acts on the sphere $X = S^n$ via the antipodal map then the equivariant coverings should correspond to the ordinary coverings of the orbit space $X/G = \mathbb{R}P^n$. Thus the elements of the covering of X can only be deformed to G-orbits. The idea of using the Lusternik-Schnirelmann category of the orbit space X/G only works nicely if the action is free. But once one leaves the realm of $\mathbb{Z}/2$-actions (or, more generally, $\mathbb{Z}/p$-actions) such an assumption is very restrictive and almost never met in applications.

Slightly generalizing the above considerations we replace the set of homogeneous G-spaces by an arbitrary set $\mathcal{A}$ of G-spaces. This means that we only consider G-coverings of X which can be deformed (equivariantly) to an element of $\mathcal{A}$. Thus the category depends on the set $\mathcal{A}$ and is denoted $\mathcal{A}$-cat. Defining not only the $\mathcal{A}$-category of a G-space but also the $\mathcal{A}$-category of a G-map we obtain as special cases the category of a pair $X' \subset X$ of G-spaces, the category of a G-pair $X' \subset X$ relative to an ambient G-space M and a generalization of the notion of genus. The genus as defined by Krasnoselski [Kr1,2] corresponds to the case $G = \mathbb{Z}/p$.

Having defined $\mathcal{A}$-cat as the basic object which we want to study in the next chapters we sketch two abstract critical point theorems and possible applications (non-linear eigenvalue problems, non-linear Dirichlet problems). These motivate the questions which we want to pursue in the chapters to follow. For instance, the non-linear eigenvalue problem leads naturally to the problem of computing the category of a sphere of a linear representation space of G.

We proceed as follows. In §2.2 we recall some notions and results from equivariant topology. Instead of proofs we give references to the literature. The category (and the genus) will be introduced in §2.3. Some basic properties will then be studied in §2.4. In §2.5 we state a critical point theorem and discuss non-linear eigenvalue problems. We are mainly interested in the question under what kind of symmetry conditions a non-linear gradient operator has an unbounded sequence of "eigenvalues". Finally, in §2.6 we state a generalization of the symmetric mountain pass theorem of Ambrosetti and Rabinowitz [AmR] to cover a more general class of symmetries. Whereas the first critical point theorem is formulated with the help of the relative category, the proof of the mountain pass theorem should illustrate the usefulness of the genus. We also sketch an application of our mountain pass theorem to the existence of infinitely many solutions of a certain non-linear elliptic system with Dirichlet boundary conditions.

2.2 Equivariant topology

In this section we collect notation and well known results on spaces with group actions which we shall use frequently. A reader who is not familiar with this material can find comprehensive introductions (which cover most of what we need) in the text books by Bredon [Bre] and tom Dieck [Di].

We fix a compact Lie group G. A *G-space* is a topological space X together with a continuous (left) action $G \times X \to X$, $(g, x) \mapsto gx$. This satisfies the usual rules: $(gh)x = g(hx)$ and $ex = x$ for all $x \in X$, $g, h \in G$, with $e \in G$ denoting the unit element. Given a closed subgroup H of G the *homogeneous G-space* G/H is the space of right cosets gH, $g \in G$. It is a G-space via multiplication from the left. A map $f: X \to Y$ between G-spaces is said to be *$(G$-)equivariant* if $f(gx) = gf(x)$. In the special case where the action of G on Y is trivial (that is $gy = y$ for all $g \in G$ and $y \in Y$), an equivariant map is called *(G)-invariant*. A *G-map* is a continuous equivariant map. The *(G)-orbit* of a point $x \in X$ is the G-subspace

$$Gx = \{gx: g \in G\}$$

of X and the *isotropy subgroup* of x is the subgroup

$$G_x = \{g \in G: gx = x\}$$

of G. A G-space X is said to be *free* if $G_x = \{e\}$ for all $x \in X$. If X is a Hausdorff space then G_x is a closed subgroup and Gx is G-homeomorphic to the homogeneous G-space G/G_x. We write

$$X^G = \{x \in X: \ gx = x \text{ for all } g \in G\}$$

for the *fixed point set* of the action and

$$X/G = \{Gx: x \in X\}$$

for the *orbit space*. X/G has the topology induced by the quotient map $X \to X/G$, $x \mapsto Gx$.

Given a closed subgroup H of G we write $NH = \{g \in G: gHg^{-1} = H\}$ for the *normalizer* of H (in G) and $WH = NH/H$ for the factor group (which is sometimes called the *Weyl group* of H in G). If X is a G-space then the fixed point subspace $X^H = \{x \in X: \ gx = x \text{ for all } g \in H\}$ is not a G-subspace in general. Instead, the restricted action of NH on X leaves X^H invariant. Therefore we have an induced action of WH on X^H.

The *conjugacy class* of H in G is the set

$$(H) = \{gHg^{-1}: g \in G\}.$$

The set $\psi(G)$ of conjugacy classes of closed subgroups is the orbit space of the G-action $(g, H) \mapsto gHg^{-1}$ defined on the set of all closed subgroups of G. If we provide the set of closed subgroups with the Hausdorff metric then $\psi(G)$ with the orbit space topology is a countable, compact Hausdorff space. A subgroup H is *subconjugate* to a subgroup K if $gHg^{-1} \subset K$ for some $g \in G$. The relation

$$(H) \leq (K) \quad \Longleftrightarrow \quad H \text{ is subconjugate to } K$$

defines a partial order on $\psi(G)$. Observe that a G-map $G/H \to G/K$ between homogeneous G-spaces exists if and only if $(H) \leq (K)$.

A *normed G-vector space* is a normed vector space V with an action of G such that each $g \in G$ acts as a linear isometry on V. In other words, the norm is G-invariant. Special classes of normed G-vector spaces are G-Banach spaces and G-Hilbert spaces. Given a G-space X, a G-Banach space V and a continuous map $f\colon X \to V$ then the map

$$f_G\colon X \longrightarrow V, \quad f_G(x) := \int_G gf(g^{-1}x)dg,$$

is equivariant. The integral is defined with the help of an invariant measure on G (a normalized Haar measure; cf. [Po]). An important special case is when V is a trivial G-space. Then f_G is G-invariant. If $f|Y$ is a G-map for some G-subspace Y of X then $f_G|Y = f|Y$. This construction can be used, for instance, to turn a metric on a G-space X into a G-invariant metric, or a norm on a G-vector space V into a G-invariant norm. We also obtain an equivariant version of the Tietze-Dugundji extension theorem, the Tietze-Gleason theorem (see [Bre], Theorem I.2.3).

2.1 Theorem:
Let A be a closed invariant subspace of the normal G-space X and let V be a G-Banach space. Then every G-map $A \to V$ can be extended to a G-map $X \to V$. $\square$

Next we define the equivariant versions of two important types of topological spaces: CW-complexes and ANRs.

2.2 Definition:
A *G-CW-complex* is a G-space X which is the union of an expanding sequence $X_0 \subset X_1 \subset X_2 \subset \ldots$ of G-subspaces X_n such that

$$X_0 \cong \coprod_{i \in I_0} G/H_i$$

is the disjoint union of homogeneous G-spaces and X_n is obtained from X_{n-1} by attaching equivariant n-cells:

$$X_n \cong X_{n-1} \cup_\varphi \left(\coprod_{i \in I_n} G/H_i \times D^n \right).$$

Here the attaching map $\varphi\colon \coprod_{i \in I_n} G/H_i \times S^{n-1} \longrightarrow X_{n-1}$ is a G-map. X has the weak (colimit) topology with respect to the filtration $(X_n\colon n \geq 0)$.

2.3 Definition:

A *G-ANR (G-equivariant absolute neighborhood retract)* is a metrizable *G*-space *X* which satisfies the following equivalent conditions:

(i) For every equivariant imbedding $i: X \to Y$ of *X* as a closed *G*-subset of a metrizable *G*-space *Y* there exists a *G*-neighborhood *U* of $i(X)$ in *Y* and a continuous equivariant retraction $U \to i(X)$.

(ii) For every closed *G*-subset *A* of a metrizable *G*-space *Y* and every *G*-map $A \to X$ there exists a continuous equivariant extension $U \to X$ to a *G*-neighborhood *U* of *A* in *Y*.

Standard references for *G*-CW-complexes are the papers by Illman [Il], Matumoto [Mat] or Waner [Wa1]. For *G*-ANRs we refer the reader to Murayama [Mur]. Examples of *G*-CW-complexes include all (finite-dimensional) smooth *G*-manifolds. Every *G*-CW-complex is a *G*-ANR. The unit sphere of a normed *G*-vector space is a *G*-ANR. More generally, paracompact Banach manifolds with a locally linear action of *G* are *G*-ANRs. (*G* acts locally linear on the Banach manifold *X* if every *G*-orbit *Gx* of *X* has a (tubular) neighborhood *U* of the form $U \cong (G \times V)/G_x$ where *V* is a G_x-Banach space and G_x acts on $G \times V$ via the diagonal action. Such an action is sometimes called locally smooth.) The following theorem is due to Murayama ([Mur], Theorem 13.3).

2.4 Theorem:
Every G-ANR has the G-homotopy type of a G-CW-complex.   □

As a consequence of this result and of the equivariant version of Whitehead's theorem (see [Mat], Theorem 5.3) we obtain the next result.

2.5 Theorem:
Let $f: X \to Y$ be a G-map between G-ANRs which induces an isomorphism

$$f_\#: \pi_n(X^H, x) \to \pi_n(Y^H, f(x))$$

for every closed subgroup H of G, every $x \in X^H$ and every $n \in \mathbb{N}$. Then f is a G-homotopy equivalence.   □

2.3 Category and genus

We first recall the general notion of equivariant category introduced by Clapp and Puppe [ClP1,2]. Fix a set $\mathcal{A}$ of *G*-spaces. A good example to keep in mind is $\mathcal{A} \subset \{G/H: H \text{ is a closed subgroup of } G\}$.

2.6 Definition:

The $\mathcal{A}$-*category* of a G-map $f\colon (X, X') \to (Y, Y')$, $\mathcal{A}$-cat(f), is the smallest integer $k \geq 0$ such that there exists a numerable covering $\{X_0, X_1, \ldots, X_k\}$ of X with the properties:

(i) $X' \subset X_0$ and there is a G-homotopy $\varphi_t\colon (X_0, X') \to (Y, Y')$, $t \in [0, 1]$, with $\varphi_0(x) = f(x)$ and $\varphi_1(x) \in Y'$ for every $x \in X_0$.

(ii) For every $i = 1, \ldots, k$ there exists $A_i \in \mathcal{A}$ and G-maps $\alpha_i\colon X_i \to A_i$, $\beta_i\colon A_i \to Y$, such that the restriction $f|X_i$ is G-homotopic to $\beta_i \circ \alpha_i$.

If no such integer exists we set $\mathcal{A}$-cat$(f) := \infty$.

2.7 Notation:

We write $\mathcal{A}$-cat(X, X') for $\mathcal{A}$-cat$(\mathrm{id}_{(X, X')})$. This is the $\mathcal{A}$-category of the G-pair (X, X'). If some ambient G-space M is fixed and (X, X') is a G-pair in M, that is $X' \subset X \subset M$, then $\mathcal{A}$-cat$_M(X, X')$ denotes the $\mathcal{A}$-category of the inclusion $(X, X') \hookrightarrow (M, X')$.

The following special case of Definition 2.6 deserves an extra name.

2.8 Definition:

The $\mathcal{A}$-*genus* of a G-space X is defined as the $\mathcal{A}$-category of the G-map $X \to$ pt: $\mathcal{A}$-genus$(X) := \mathcal{A}$-cat$(X \to \mathrm{pt})$.

Let us comment on these notions. Intuitively, $\mathcal{A}$-cat(X, X') means: Take out the largest G-subset X_0 of X containing X' which can be deformed into X' in X; X_0 must be empty if $X' = \emptyset$. Then cover the rest by G-subsets $X_1, \ldots, X_k$ each of which can be deformed into some "$\mathcal{A}$-point" $A_i \to X$ in X. In the important special case $\mathcal{A} = \{G/H\colon H$ is a closed subgroup of $G\}$ the $\mathcal{A}$-category is the obvious equivariant version of the usual Lusternik-Schnirelmann category. The homogeneous spaces G/H should be thought of as "equivariant points". The definition of $\mathcal{A}$-cat$_M(X)$ differs from the one of $\mathcal{A}$-cat(X) in that the deformations are allowed to take place in M and not just in X. This notion is particularly suited for applications to critical point theory.

The $\mathcal{A}$-genus is a natural generalization of the usual genus which is defined for spaces with an involution (an action of $\mathbb{Z}/2$). Clearly, $\mathcal{A}$-genus(X) is the smallest integer $k \geq 0$ such that there exists a numerable covering $\{X_1, \ldots, X_k\}$ of X, elements $A_i \in \mathcal{A}$ and G-maps $X_i \to A_i$ for $i = 1, \ldots, k$. For another equivalent description we need to recall the *join* of G-spaces $X_1, \ldots, X_k$:

$$X_1 * \ldots * X_k := \left\{ [s_1, x_1, \ldots, s_k, x_k]\colon s_i \in [0, 1],\ \sum_{i=1}^{k} s_i = 1,\ x_i \in X_i \right\}.$$

Here $[s_1, x_1, \ldots, s_k, x_k] = [t_1, y_1, \ldots, t_k, y_k]$ if and only if for every $i = 1, \ldots, k$: $s_i = t_i$ and $x_i = y_i$ or $s_i = t_i = 0$. This is again a G-space with the action

$$g[s_1, x_1, \ldots, s_k, x_k] = [s_1, gx_1, \ldots, s_k, gx_k].$$

If the $X_i = SV_i$ are unit spheres of normed G-vector spaces V_i then

$$X_1 * \ldots * X_k \cong S(V_1 \times \ldots \times V_k).$$

More generally, for G-subsets $X_i \subset SV_i$ one may consider $X_1 * \ldots * X_k$ as a G-subset of $S(V_1 \times \ldots \times V_k)$ via the identification

$$[s_1, x_1, \ldots, s_k, x_k] \longmapsto \rho(s_1 x_1, \ldots, s_k x_k)$$

where $\rho: V - 0 \to SV$ is the radial retraction.

2.9 Proposition:
$\mathcal{A}$-genus(X) is the smallest integer $k \geq 0$ such that there exist $A_1, \ldots, A_k \in \mathcal{A}$ and a G-map $X \longrightarrow A_1 * \ldots * A_k$.

Proof:
Set $\gamma := \mathcal{A}$-genus(X) and let k be the smallest integer as above. In order to see $\gamma \leq k$ consider a G-map $X \longrightarrow A_1 * \ldots * A_k$. Then $X = X_1 \cup \ldots \cup X_k$ with

$$X_i := \{x \in X: f(x) = [s_1, a_1, \ldots, s_k, a_k] \text{ and } s_i > 0\}.$$

This is a numerable equivariant covering because the maps $X \to [0,1]$, $x \mapsto s_i$ if $f(x) = [s_1, a_1, \ldots, s_k, a_k]$, provide a G-invariant partition of unity subordinated to $\{X_1, \ldots, X_k\}$. Moreover, for any $i = 1, \ldots, k$ we set $\alpha_i: X_i \to A_i$, $x \mapsto a_i$ if $f(x) = [s_1, a_1, \ldots, s_k, a_k]$. Thus $\gamma = \mathcal{A}$-genus$(X) = \mathcal{A}$-cat$(X \to \mathrm{pt}) \leq k$.

To see the inverse inequality let $\{X_1, \ldots, X_\gamma\}$ be a covering of X with subordinated partition of unity $\pi_i: X \to [0,1]$. We also have G-maps $X_i \to A_i$ with $A_i \in \mathcal{A}$. Then we can define a G-map $f: X \to A_1 * \ldots * A_\gamma$ by setting

$$f(x) := [\pi_1(x), \alpha_1(x), \ldots, \pi_\gamma(x), \alpha_\gamma(x)].$$

It does not matter that α_i is not defined outside of X_i since $\pi_i(X - X_i) = 0$. This shows $k \leq \gamma = \mathcal{A}$-genus$(X)$. $\qquad\square$

If $G = \mathbb{Z}/2$ and $\mathcal{A} = \{G\}$ then the join of k copies of G is G-homeomorphic to S^{k-1} where G acts on S^{k-1} via the antipodal map. In this case (and also if $G = \mathbb{Z}/p$) we arrive at the usual notion of genus due to Krasnoselski [Kr1,2] and Yang [Y2] (in [Y2] it is called B-index); see also the paper [Cof] of Coffman for a version similar to the one in Proposition 2.9. The genus is obviously a lower bound for the category.

2.10 Proposition:
For any G-space X: $\mathcal{A}$-genus$(X) \leq \mathcal{A}$-cat(X). $\qquad\square$

It is not difficult to find examples where the strict inequality holds. A trivial one is the case where $X = G \sqcup G$ so that $\mathcal{A}$-genus$(X) = 1$ and $\mathcal{A}$-cat$(X) \geq 2$ independently of $\mathcal{A}$. Another trivial example can be obtained when $\mathcal{A}$ contains an element A with $A^G \neq \emptyset$. Then $\mathcal{A}$-genus$(X) = 1$ for every non-empty G-space X. Some less obvious examples for the strict inequality can be found in chapter 3. There we shall construct fixed point free G-spaces X such that X^H is connected for every subgroup H of G but $\mathcal{A}$-cat$(X) = \infty$ and $\mathcal{A}$-genus$(X) < \infty$. Here $\mathcal{A} = \{G/H\colon H \subsetneqq G$ a closed subgroup$\}$. The importance of the genus lies mainly in its computability. In order to show the inequality $\mathcal{A}$-genus$(X) \geq k$ one has to prove the non-existence of a G-map $X \to A_1 * \ldots * A_{k-1}$. This can often be done using algebraic topology.

2.11 Remark:

As mentioned in the beginning of this section the $\mathcal{A}$-category as defined in 2.6 is due to Clapp and Puppe [ClP1,2]. An important special case has been defined earlier by Fadell [Fa2]; see also [Ram] for various generalizations. More precisely, Fadell introduced G-cat$(X)= \mathcal{A}$-cat(X) with

$$\mathcal{A} = \{G/H\colon H \subset G \text{ a closed subgroup}\}.$$

In the non-equivariant situation several versions of a relative category have been introduced first by Reeken [Re1,2] and, later, by Fadell [Fa3] and Fournier and Willem [FoW]. The $\mathcal{A}$-genus(X) with $\mathcal{A} = \{G\}$ is due to Yang [Y2] for $G = \mathbb{Z}/2$ under the name B-index, to Krasnoselski [Kr1,2] for $G = \mathbb{Z}/p$, to Švarc [Sv] for free actions of discrete groups, and for arbitrary G acting freely on X to Fadell [Fa1]. In the case $G = S^1$ Benci introduced the $\mathcal{A}$-genus (he called it *geometrical index*) with $\mathcal{A} = \{G/H\colon H \subsetneqq G$ a closed subgroup$\}$. This case is particularly important for applications. In [Ba2] the author introduced $\mathcal{A}$-genus for arbitrary G and $\mathcal{A}$ as above. There one can also find lower bounds for $\mathcal{A}$-genus(SV) for representations V of the cyclic group $G = \mathbb{Z}/n$ as well as for the $\mathbb{Z}/2$-genus of lens spaces. It is also possible to define relative versions of the genus; see for example [Sz2], Section 2, or [BC2], Remark 2.8b). In [Ko2] Komiya introduced the set-valued genus $\alpha(X)$ of a G-space X. In the situation considered here this is the set of all joins $A_1 * \ldots * A_k$ with $A_i \in \mathcal{A}$ and such that there exists a G-map $X \to A_1 * \ldots * A_k$.

Our version of the $\mathcal{A}$-category of a G-map is very general and encompasses various other definitions of relative and/or equivariant Lusternik-Schnirelmann category. But for some applications one needs an even more general concept where one also specifies and restricts the class of G-homotopies in Definition 2.6. Possible generalizations in this direction have been introduced by Szulkin [Sz3] and Fournier et al. [FLRW] in a non-equivariant context and by Bartsch and Clapp [BC2] within the equivariant world. In these papers the authors had to deal with strongly indefinite functionals. The category as defined in 2.6 cannot be applied to these situations. It would be too small and would yield only trivial results. In [Sz3] and [BC2] the $(G$-$)$homotopies have to satisfy additional compactness and boundedness conditions so that they can be approximated by finite-dimensional maps whereas in [FLRW] all spaces and maps come together with finite-dimensional approximations (Galerkin method). This corresponds to the fact that it is not possible to extend the concept of Brouwer degree to the infinite-dimensional setting without restricting the class of maps. Of course, this makes it more difficult to prove a deformation lemma needed for applications to critical point theory. It is clear that the class of homotopies entering the definition of the category has to correspond to the deformations appearing in the application.

2.4 Properties of category and genus

Here we collect a number of properties of $\mathcal{A}$-category (and $\mathcal{A}$-genus) which are important for applications to critical point theory. Since most of these properties are variations of known results about the Lusternik-Schnirelmann category and the $\mathbb{Z}/2$-genus we only give sketch-proofs or refer to the literature.

2.12 Proposition:

Fix a G-space M. Then $\mathcal{A}$-cat_M has the following properties.

Normalization: *$\mathcal{A}$-$\mathrm{cat}_M(A) = 1$ for any G-subset A of M which is G-homeomorphic to an element of $\mathcal{A}$. And $\mathcal{A}$-$\mathrm{cat}_M(X,X) = 0$ for any G-subset X of M.*

Deformation monotonicity: *Let X, Y and Z be G-subsets of M with $Z \subset X \cap Y$ and suppose there exists a G-homotopy $h_t \colon (X, Z) \longrightarrow (M, Z)$, $t \in [0, 1]$, with $h_0(x) = x$ and $h_1(x) \in Y$ for all $x \in X$. Then*

$$\mathcal{A}\text{-}\mathrm{cat}_M(X, Z) \le \mathcal{A}\text{-}\mathrm{cat}_M(Y, Z).$$

Subadditivity: *Suppose M and all elements of $\mathcal{A}$ are G-ANRs. If $Z \subset X$ and Y are closed G-subsets of M with $Y \cap Z = \emptyset$ then*

$$\mathcal{A}\text{-}\mathrm{cat}_M(X \cup Y, Z) \le \mathcal{A}\text{-}\mathrm{cat}_M(X, Z) + \mathcal{A}\text{-}\mathrm{cat}_M(Y).$$

Continuity: *Suppose M and all elements of $\mathcal{A}$ are G-ANRs. Then every closed G-subset X of M has a G-neighborhood U in M with $\mathcal{A}$-$\mathrm{cat}_M(U) = \mathcal{A}$-$\mathrm{cat}_M(X)$.*

Moreover, if M and all elements of $\mathcal{A}$ are G-ANRs then $\mathcal{A}$-cat_M is maximal with respect to the first three properties.

Proof:

The proof is similar to the one of Proposition 3.4 in [ClP2] (cf. also [ClP1] for the continuity property), although the authors work with a slightly different version of the relative category. The maximality of $\mathcal{A}$-cat_M is easy: Suppose we have a function γ which associates to a G-pair (X, X') of M an integer $\gamma(X, X') \ge 0$ satisfying the above properties. We claim that $\gamma(X, X') \le \mathcal{A}$-$\mathrm{cat}_M(X, X') =: k$. This can be seen as follows. By definition $X = X_0 \cup X_1 \cup \ldots \cup X_k$ as in 2.6. We may also assume that $X_i \cap X' = \emptyset$ for $i = 1, \ldots, k$. Then $\gamma(X_0, X') = 0$ by deformation monotonicity and normalization. For the same reasons $\gamma(X_i) = 1$, hence the subadditivity property yields:

$$\gamma(X, X') \le \gamma(X_0, X') + \gamma(X_1) + \ldots + \gamma(X_k) \le k.$$

$\square$

It is easy to see that one cannot replace $\mathcal{A}$-cat$_M$ by $\mathcal{A}$-cat in Proposition 2.12. This is true already in the non-equivariant situation.

Next we want to find an upper bound for the $\mathcal{A}$-category of a G-space X. In the non-equivariant case this is provided by the dimension of X (cf. [P2], Theorem 6.4). In order to avoid the trivial case where $\mathcal{A}$-cat$(X) = \infty$ simply because there exists a G-orbit Gx in X which cannot be mapped equivariantly into any element A of $\mathcal{A}$ we work with the following sets:

$\mathcal{A}_X := \{G/G_x : x \in X\}$ where X is a G-space;

$\widehat{\mathcal{A}} := \{A_1 \sqcup \ldots \sqcup A_k : k \geq 1,\ A_i \in \mathcal{A}\}$ where $\mathcal{A}$ is any set of G-spaces.

2.13 Proposition:
For a compact G-ANR X the following holds:

a) $\mathcal{A}_X$-cat$(X) < \infty$.

b) $\widehat{\mathcal{A}}_X$-cat$(X) \leq 1 + \dim(X/G)$.

Proof:
a) Since X is a G-ANR every orbit Gx has an invariant G-neighborhood $U(x)$ which can be deformed to Gx inside X. Finitely many such neighborhoods suffice to cover X because X is compact.

b) Suppose $n := \dim(X/G) < \infty$. For every orbit Gx in X choose an open G-neighborhood $U(x)$ as above. The sets $U(x)/G \subset X/G$ cover X/G. Lemma 2.4 of [P1] yields a locally finite open covering $\{V_{i,\beta} : i \in \{1,\ldots,n+1\},\ \beta \in B_i\}$ refining the covering $\{U(x)/G : x \in X\}$ and such that $V_{i,\beta} \cap V_{i,\beta'} = \emptyset$ if $\beta \neq \beta'$. Since X is compact we may assume that B_i is finite for $i = 1,\ldots,n+1$. Then

$$W_i := \bigcup_{\beta \in B_i} V_{i,\beta}\,, \quad i = 1,\ldots,n+1,$$

can be deformed inside X to an element of $\widehat{\mathcal{A}}_X$. Obviously, $X = \bigcup_{i=1}^{n+1} W_i$, so that $\widehat{\mathcal{A}}_X$-cat$(X) \leq 1 + n$.   □

It is not difficult to replace X in Proposition 2.13 by a G-map. One possible generalization goes as follows: If $f : X \to Y$ is a G-map such that $f(X)$ is compact and Y is a G-ANR then $\mathcal{A}_X$-cat$(f) < \infty$. But other combinations of a compactness condition and a G-ANR condition work equally well (even in the relative case).

The formula in 2.13b) can be wrong if X is not compact. For a non-compact but paracompact G-ANR X one has to take the (cardinal) number $c(H)$ of components of X^H/NH, $H \subset G$, into account. If this is infinite for some $H \subset G$ then $\widehat{\mathcal{A}}_X$-cat$(X) = \infty$. In that case we choose an index set I such that card$(I) \geq c(H)$ for all closed subgroups H of G. Then $\mathcal{A}_X^I$-cat$(X) \leq 1 + \dim(X/G)$ where

$$\mathcal{A}_X^I := \left\{ \coprod_{i \in J} A_i : J \subset I,\ A_i \in \mathcal{A}_X \right\}.$$

It is also clear that 2.13b) is false if we replace $\widehat{\mathcal{A}}_X$ by $\mathcal{A}_X$ because X^H/NH need not be connected in general. Again, one has to take the numbers $c(H)$ into account. It is not difficult to prove that

$$\mathcal{A}_X\text{-cat}(X) \leq \left(1 + \dim(X/G)\right) \cdot \max_{H \subset G} c(H).$$

Of course, this is only of interest if $\max c(H) < \infty$.

The next result is also quite easy to prove. We consider the following situation. Given a compact Lie group G and a closed subgroup H of G. Then any G-space X can be considered as an H-space by forgetting part of the action. It happens relatively often that the category of the G-space X is more difficult to compute than the category of X considered as an H-space. This will be the case for example in the applications studied in chapters 8 and 9.

2.14 Proposition:
Let H be a closed subgroup of G. Consider a set $\mathcal{A}$ of G-spaces and a set $\mathcal{B}$ of H-spaces. Then for any G-map $f\colon (X, X') \longrightarrow (Y, Y')$:

$$\mathcal{B}\text{-cat}(f) \leq \mathcal{A}\text{-cat}(f) \cdot \max\{\mathcal{B}\text{-cat}(A)\colon A \in \mathcal{A}\}.$$

Proof:
Set $k = \mathcal{A}\text{-cat}(f)$ and $\mu = \max\{\mathcal{B}\text{-cat}(A)\colon A \in \mathcal{A}\}$. Let $\{X_0, X_1, \ldots, X_k\}$ be a G-covering of X and $\alpha_i\colon X_i \to A_i$, $i = 1, \ldots, k$, be the G-maps according to the definition of $\mathcal{A}\text{-cat}(f)$. Each A_i can be covered by $\mu_i \leq \mu$ open H-subspaces $A_{i,j}$ which can be deformed (inside A_i) H-equivariantly into some $B_{i,j} \in \mathcal{B}$. Then the covering $\{\alpha_i^{-1}(A_{i,j})\colon i = 1, \ldots, k, \ j = 1, \ldots, \mu_i\}$ is an H-covering of X as required in Definition 2.6.   □

If $\mathcal{A}$ is a set of homogeneous G-spaces and $\mathcal{B}$ a set of homogeneous H-spaces it would be interesting to compute $\mathcal{B}\text{-cat}(A)$ for $A \in \mathcal{A}$. If $A = G/K$ then a lower bound for $\mathcal{B}\text{-cat}(G/K)$ is $\text{cat}(H \backslash G/K)$. Here $H \backslash G/K = (G/K)/H$ is a double coset space. An important special case is when G is connected and H is the maximal torus of G. Some results concerning the category of homogeneous spaces have been obtained by Singhof [Si1,2]. I am not aware of computations of the category of double coset spaces. With respect to Proposition 2.14 it would be most interesting to have good upper bounds for $\mathcal{B}\text{-cat}(A)$.

Now we collect similar results for $\mathcal{A}$-genus.

2.15 Proposition:
$\mathcal{A}$-genus has the following properties.

Normalization: *$\mathcal{A}\text{-genus}(A) = 1$ for every $A \in \mathcal{A}$ and $\mathcal{A}\text{-genus}(X) = 0$ if and only if $X = \emptyset$.*

Monotonicity: *If there exists a G-map $X \to Y$ then $\mathcal{A}\text{-genus}(X) \leq \mathcal{A}\text{-genus}(Y)$.*

Subadditivity: *If X_1 and X_2 are invariant subspaces of the normal G-space X whose interiors cover X then*

$$\mathcal{A}\text{-genus}(X_1 \cup X_2) \leq \mathcal{A}\text{-genus}(X_1) + \mathcal{A}\text{-genus}(X_2).$$

Continuity: *Suppose all elements of $\mathcal{A}$ are G-ANRs. Then every closed G-subset Y of a metrizable G-space X has an invariant neighborhood U such that*

$$\mathcal{A}\text{-genus}(U) = \mathcal{A}\text{-genus}(Y).$$

Moreover, $\mathcal{A}$-genus is maximal with respect to the first three properties.

Proof:

We use the description of $\mathcal{A}$-genus given in Proposition 2.9. Normalization and monotonicity are obvious. To prove subaddititvity we construct a G-invariant partition of unity, $\{\pi_1, \pi_2\}$, subordinate to $\{X_1, X_2\}$. Given G-maps $f_i \colon X_i \to Y_i$, $i = 1, 2$, where Y_i is a join of $\mathcal{A}$-genus(X_i) elements of $\mathcal{A}$ we define a new G-map $f \colon X \to Y_1 * Y_2$ by setting $f(x) := [\pi_1(x), f_1(x), \pi_2(x), f_2(x)]$. This is well defined because $\pi_i(x) = 0$ if $x \notin X_i$. This shows the subadditivity property. Continuity is an easy consequence of Definition 2.3(ii) because the join of G-ANRs is again a G-ANR. Finally, consider a function γ which satisfies the normalization, monotonicity and subadditivity properties. Then $\gamma \leq \mathcal{A}$-genus follows from the fact that $\gamma(A_1 * \ldots * A_k) \leq k$ if all A_i are elements of $\mathcal{A}$. The latter inequality is clear because the sets

$$U_i = \{[s_1, a_1, \ldots, s_k, a_k] \in A_1 * \ldots * A_k \colon s_i > 0\}, \quad i = 1, \ldots, k,$$

can be mapped equivariantly to A_i, so $\gamma(U_i) = 1$. Obviously, $A_1 * \ldots * A_k = \bigcup_{i=1}^{k} U_i$, hence, $\gamma(A_1 * \ldots * A_k) \leq k$. □

2.16 Proposition:

For a compact G-space X the following holds:

a) *$\mathcal{A}$-genus$(X) < \infty$ if each orbit of X can be mapped equivariantly into some element of $\mathcal{A}$. This is the case for instance if $\mathcal{A} = \mathcal{A}_X$.*

b) *$\mathcal{A}$-genus$(X) \leq 1 + \dim(X/G)$ if each finite disjoint union of orbits of X can be mapped into some element of $\mathcal{A}$. This is the case for instance if $\mathcal{A} = \widehat{\mathcal{A}}_X$.*

Proof:

Since a compact space is completely regular every orbit Gx has a G-neighborhood $U(x)$ which retracts equivariantly onto Gx. This is a consequence of the existence of a tube about an orbit (see [Bre], Theorem II.5.4). Now one can argue as in the proof of Proposition 2.13. □

Observe that we did not need that X is a G-ANR as in 2.12. If X is a paracompact G-space then $\mathcal{A}_X^{\mathbb{N}}$-genus$(X) \leq 1 + \dim(X/G)$ because there exist only countably many homogeneous G-spaces (up to G-homeomorphism). Therefore every disjoint union $\coprod_{i \in I} A_i$ can be mapped equivariantly into a countable disjoint union.

2.17 Proposition:
  Let H be a closed subgroup of G. Consider a set $\mathcal{A}$ of G-spaces and a set $\mathcal{B}$ of H-spaces. Then for any G-space X:

$$\mathcal{B}\text{-genus}(X) \leq \mathcal{A}\text{-genus}(X) \cdot \max\{\mathcal{B}\text{-genus}(A) \colon A \in \mathcal{A}\}.$$

□

The monotonicity property of the genus as formulated in Proposition 2.15 is false for $\mathcal{A}$-cat and $\mathcal{A}$-cat$_M$. Trivial counterexamples can be constructed with $Y = $ pt the one point space. But the following weaker version of monotonicity still holds (see [BCP], Lemma (4.3)).

2.18 Proposition:
  Weak Monotonicity: *If there exists a G-map $X \to Y$ then*

$$\mathcal{A}\text{-cat}(X * Y) \leq 2 \cdot \mathcal{A}\text{-cat}(Y).$$

Proof:
  X and Y can be considered as G-subspaces of $X * Y$. The natural inclusion $X \hookrightarrow X * Y - Y$ is a G-homotopy equivalence with inverse

$$\pi_X \colon X * Y - Y \longrightarrow X, \quad [s, x, 1 - s, y] \longmapsto x;$$

similarly for Y. If $\{Y_1, \ldots, Y_k\}$ is a numerable covering of Y according to the definition of $k = \mathcal{A}\text{-cat}(Y)$ then $X * Y - X$ is covered by $\{\pi_Y^{-1}(Y_i) \colon i = 1, \ldots, k\}$. It is easy to see that $\pi_Y^{-1}(Y_i)$ can be deformed to an element A_i of $\mathcal{A}$ because one can first deform it into $Y_i \subset Y \subset X * Y$ and then deform Y_i to A_i. Next, $X * Y - Y$ is covered by $\{(f \circ \pi_X)^{-1}(Y_i) \colon i = 1, \ldots, k\}$; here $f \colon X \to Y$ is any G-map. As above, one first deforms $(f \circ \pi_X)^{-1}(Y_i)$ to $f^{-1}(Y_i) \subset X \subset X * Y$. Then one applies the G-homotopy

$$f^{-1}(Y_i) \times [0, 1] \longrightarrow X * Y,$$
$$([1, x, 0, *], t) \longmapsto [1 - t, x, t, \alpha_i(f(x))].$$

Here $\alpha_i \colon Y_i \to A_i$ is a G-map as in Definition 2.6. This shows that $\mathcal{A}\text{-cat}(X * Y) \leq 2k$ as claimed.   □

2.5 Critical point theory with symmetries

This section has mainly a motivating character. We first state an abstract critical point theorem. Let M be a paracompact C^{2-}-Banach manifold and assume that M is a G-ANR. Moreover we assume that multiplication with an element g of G is a C^1-map $M \to M$ and that the induced map

$$G \times TM \longrightarrow TM, \quad (g,v) \longmapsto dg\big(\pi(v)\big)v,$$

is continuous and locally Lipschitz continuous in the second variable. Suppose there exists a Finsler structure $\| - \| : TM \to \mathbb{R}$ for the tangent bundle $\pi : TM \to M$ such that M is complete with respect to the associated Finsler metric (see [P2], §§2 and 3). We are interested in critical points of a G-invariant C^1-function $\Phi : M \to \mathbb{R}$ with associated critical values in an interval $(a,b] \subset \mathbb{R}$, $a,b \in \mathbb{R}$. Observe that a critical point $x \in M$ gives rise to a G-orbit Gx of critical points, a critical G-orbit. We want to find a lower estimate for the number of critical G-orbits in $\Phi^{-1}(a,b]$. We write Φ^c for the set of all $x \in M$ with $\Phi(x) \leq c$ and $K_c = \{x \in M : \Phi'(x) = 0, \ \Phi(x) = c\}$ for the set of critical points at the level c.

2.19 Theorem:

  Suppose a is not a critical value of Φ. Then Φ has at least $\mathcal{A}$-cat$_M(\Phi^b, \Phi^a)$ different critical G-orbits in $\Phi^{-1}(a,b]$ provided the following two conditions hold:

(Φ_1) *Φ satisfies the Palais-Smale condition in $[a,b]$, i.e. any sequence (x_n) in $\Phi^{-1}[a,b]$ with $\|\Phi'(x_n)\| \to 0$ as $n \to \infty$ has a convergent subsequence.*

(Φ_2) *$\mathcal{A}$ contains up to G-homeomorphism all critical G-orbits in $\Phi^{-1}(a,b]$.*

More precisely, under these conditions the following holds. The function

$$m : [a,b] \to \mathbb{N}, \quad c \mapsto \mathcal{A}\text{-cat}_M\big(\Phi^c, \Phi^a\big)$$

is weakly increasing and has the following properties:

(i) *There is an $\epsilon > 0$ such that $m(a + \epsilon) = 0$.*

(ii) *For every $c \in (a,b]$ there is an $\epsilon > 0$ such that*

$$m(c + \epsilon) \leq m(c - \epsilon) + \mathcal{A}\text{-cat}_M(K_c).$$

(iii) *If the set of critical values is bounded by b then $m(b) = \mathcal{A}\text{-cat}_M(M, \Phi^a)$.*

Proof:

  The proof follows standard lines. One first proves an equivariant deformation lemma by integrating an equivariant pseudo-gradient vector field. Then one applies the properties of $\mathcal{A}$-cat$_M$ as stated in Proposition 2.12; see for instance [ClP2], Theorem 3.5. $\square$

The theorem extends to the case $b = \infty$, that is $\Phi^b = M$. If $\mathcal{A}\text{-cat}_M(M, \Phi^a) = \infty$ then there must exist an unbounded sequence $c_n \to \infty$ of critical values. Moreover, if Φ is bounded from below and satisfies the Palais-Smale condition everywhere then it has at least $\mathcal{A}\text{-cat}(M)$ critical G-orbits.

It is also possible to give a minimax characterization of the critical values. Set

$$S_i := \{X \subset \Phi^b : \mathcal{A}\text{-cat}_M(X, X \cap \Phi^a) \geq i\}.$$

It is easy to see that the numbers

$$c_i := \inf_{X \in S_i} \sup \left(\Phi(X) \right), \quad i = 1, \dots, \mathcal{A}\text{-cat}_M\left(\Phi^b, \Phi^a \right),$$

are critical values in $(a, b]$. If $c = c_i = c_{i+1} = \dots = c_{i+k}$ then the set K_c of critical values at level c satisfies $\mathcal{A}\text{-cat}_M(K_c) \geq k + 1$.

An interesting non-standard variation of Theorem 2.19 concerns the case when M is only a C^1-manifold. Such a situation occurs for instance when one has to deal with elliptic problems which are not quadratic. In the non-equivariant case critical point theory on C^1-manifolds has been studied by various authors; see the papers [Sz1] by Szulkin or [CorDM] by Corvellec et al. and the references therein. It is possible to extend Theorem 2.19 to the case of C^1-manifolds M assuming on the other hand that the action of G on M is locally linear (M is then automatically a G-ANR). If $\Phi^a = \emptyset$ one can essentially follow the proof in [Sz1]. In the general case one can use the approach of [CorDM].

A simple and well known situation covered by Theorem 2.19 is the (non-equivariant) mountain pass theorem. There M is a Banach space and Φ^a is not connected, hence $\mathcal{A}\text{-cat}_M(M, \Phi^a) \geq 1$. If $G = \mathbb{Z}/2$ acts on M via the antipodal map then a symmetric version of the mountain pass theorem has been proved by Ambrosetti and Rabinowitz [AmR]; cf. also [Ra2]. In this situation one can prove that $\mathcal{A}\text{-cat}_M(M, \Phi^a) = \infty$ which guarantees the existence of an unbounded sequence of critical values. This approach provides another proof of the $\mathbb{Z}/2$-equivariant mountain pass theorem which also works for more general group actions (see [FaHR] for the case $G = S^1$, [ClP2] for the case of a torus or a p-torus and [BCP] for the case where G is an extension of a torus by a finite p-group). Another generalization to a very general class of group actions not covered by any of the above papers will be proved in the next section together with an application to a non-linear elliptic system with Dirichlet boundary conditions.

A special case of Theorem 2.19 which is important for applications is when $M = SX$ is the unit sphere in a (possibly infinite-dimensional) G-Hilbert space X. Then critical points of the G-invariant functional $\Phi : M \to \mathbb{R}$ are solutions of the non-linear eigenvalue problem

$$(*) \qquad \nabla\Phi(x) = \lambda \nabla\Psi(x)$$

where $\Psi(x) = \frac{1}{2}\|x\|^2$ and λ occurs as Lagrange multiplier. For various applications it is of interest to look at other G-invariant functionals $\Psi : X \to \mathbb{R}$ and $M := \Psi^{-1}(1)$. We refer the reader to [Bro], [Z] and the references therein for a discussion of non-linear eigenvalue problems. In [Bro] the importance of group invariance is emphasized, but only the case of free actions is treated. Suppose for simplicity that Φ is bounded from below and satisfies the assumptions of Theorem 2.19 with $a < \inf \Phi$ and $b = \infty$. Then $\mathcal{A}\text{-cat}_M(M)$ is a lower bound for the number of critical G-orbits of Φ. If $G = (\mathbb{Z}/p)^k$ is a p-torus and $\mathcal{A} = \{G/H \colon H \subsetneq G\}$ then

$$\mathcal{A}\text{-cat}(SX) = \mathcal{A}\text{-genus}(SX) = \begin{cases} \dim X & \text{if } X^G = 0; \\ \infty & \text{if } X^G \neq 0. \end{cases}$$

Similarly for $G = (S^1)^k$ a torus and $\mathcal{A} = \{G/H \colon H \subsetneq G \text{ a closed subgroup}\}$:

$$\mathcal{A}\text{-cat}(SX) = \mathcal{A}\text{-genus}(SX) = \begin{cases} \tfrac{1}{2}\dim X & \text{if } X^G = 0; \\ \infty & \text{if } X^G \neq 0. \end{cases}$$

Replacing $\mathcal{A}$ by $\mathcal{A}_X$ we obtain the same result in the case $X^G = 0$, whereas

$$1 = \mathcal{A}_X\text{-genus}(SX) \leq \mathcal{A}_X\text{-cat}(SX) \leq 2 \qquad \text{if } X^G \neq 0.$$

As a consequence, for these groups the infinite-dimensional eigenvalue problem $\nabla\Phi(x) = \lambda x$ has infinitely many solutions if $\Phi \colon X \to \mathbb{R}$ is G-invariant, bounded from below and satisfies the Palais-Smale condition (provided $X^G = 0$, of course). The sequence of eigenvalues $\lambda_1 \leq \lambda_2 \leq \ldots$ is unbounded as one expects. These results are well known. An elementary proof using only the concept of degree and Sard's lemma can be found in [Ba4]. They are also a consequence of the theorems 5.1–5.3 below (cf. Remark 5.6). The following stronger result is more difficult to prove. It is due to Clapp and Puppe [ClP2], Proposition 6.3. A relative version can be found in [BC2], Theorem A.1.

2.20 Theorem:

Let X be a (possibly infinite-dimensional) representation of the (p-)torus G with $X^G = 0$. Then

$$\widehat{\mathcal{A}}\text{-cat}(SX) = \widehat{\mathcal{A}}\text{-genus}(SX) = \begin{cases} \dim X & \text{if } G = (\mathbb{Z}/p)^k \text{ is a p-torus;} \\ \tfrac{1}{2}\dim X & \text{if } G = (S^1)^k \text{ is a torus.} \end{cases}$$

Here $\widehat{\mathcal{A}}$ contains (up to G-homeomorphism) all fixed point free G-spaces whose orbit spaces are finite (and discrete).   □

Combining this theorem with the results of §2.4 we obtain the following lower estimate for category and genus.

2.21 Corollary:

Consider an infinite compact Lie group G with maximal torus T. Let X be a (possibly infinite-dimensional) representation of G such that $X^T = 0$. Then

$$\widehat{\mathcal{A}}_{SX}\text{-cat}(SX) \geq \widehat{\mathcal{A}}_{SX}\text{-genus}(SX) \geq (\dim X)/2(1 + \dim G - \dim T).$$

Proof:

Let $\mathcal{B}$ be the set of homogeneous T-spaces different from T/T. Then Theorem 2.20 and Proposition 2.17 imply

$$\widehat{\mathcal{A}}_{SX}\text{-genus}(SX) \geq (\dim X)/2 \max \left\{ \widehat{\mathcal{B}}\text{-genus}(A) \colon A \in \widehat{\mathcal{A}}_{SX} \right\}.$$

Now

$$\widehat{\mathcal{B}}\text{-genus}(A) \leq 1 + \dim(A/T) \leq 1 + \dim G - \dim T$$

according to Proposition 2.16b). $\qquad\qquad\qquad\qquad\qquad\qquad\qquad\qquad\qquad$ $\square$

It is natural to ask whether Theorem 2.20 can be generalized to other groups and whether the restriction $X^T = 0$ in Corollary 2.21 is really necessary. This leads to the following set of questions.

2.22 Problems:

Is it true that $\mathcal{A}_{SX}\text{-cat}(SX) = \infty$ if X is an infinite-dimensional representation of the compact Lie group G with $X^G = 0$? May be there exists even a number $a_G > 0$ such that $\mathcal{A}_{SX}\text{-cat}(SX) \geq a_G \cdot \dim X$ for representations X of G with $X^G = 0$. Being very optimistic, why shouldn't we try $a_G = 1$ for finite G? After all, this works for p-tori. If this is not the case then compute $\mathcal{A}_{SX}\text{-cat}(SX)$ or find at least interesting lower bounds for it! Of course, we can ask the same questions for the genus instead of the category.

We address these problems in chapters 3 and 5. Our main results are stated in §§3.2 and 3.5. We want to emphasize that these questions are not only interesting for functionals $\Phi \colon M \to \mathbb{R}$ with $M = SX$ and $M^G = \emptyset$. If $M^G \neq \emptyset$ we set $a := \sup \Phi(M^G)$. Then one can try to find critical points above the level a. Theorem 2.19 says that we have to compute $\mathcal{A}\text{-cat}_M(M, \Phi^a)$. This number can be estimated from below by $\mathcal{A}\text{-cat}_M(M, M^G)$. If $M = SX$ is the sphere in a G-Hilbert space X then this is closely related to $\mathcal{A}\text{-cat}(S(X^G)^\perp)$. In [BC2] the notion of *compact equivariant category*, $\mathcal{A}\text{-cat}^c$, has been developed and applied to the strongly indefinite Hamiltonian action functional in order to obtain multiple periodic solutions of symmetric Hamiltonian systems. There it is (implicitly) proved that

$$\mathcal{A}\text{-cat}^c(SX, SY) = \mathcal{A}\text{-cat}(SY^\perp)$$

if G is a torus. Here Y is a G-invariant closed linear subspace of X. It is allowed that $\dim Y = \infty$ and $\dim Y^\perp = \infty$. Moreover, it is possible to replace SX by a G-hypersurface M which is radially G-homeomorphic to SX but which is unbounded. (In [BC2] M is the boundary of such an unbounded starshaped neighborhood of 0 and SY is replaced by $Y \cap M$.)

The non-linear eigenvalue problem $(*)$ also leads to another set of questions.

2.23 Problems:

Consider again the non-linear eigenvalue problem

$(*)$ $$\qquad\qquad\qquad\qquad \nabla\Phi(x) = \lambda\nabla\Psi(x) = \lambda x$$

and suppose that $\nabla\Phi(0) = 0$. This time we are interested in solutions (λ, x) of $(*)$ which are near $(\lambda_0, 0)$, i.e. we want to find solutions which bifurcate from the given branch $\mathbb{R} \times \{0\}$. When one thinks of λ as a parameter then it is natural to ask whether for any λ close to λ_0 there exist solutions (λ, x) of $(*)$ with x close to 0. This question makes sense (and is of interest) in a more general setting. Consider a continuous family $\Phi_\lambda : X \rightarrow \mathbb{R}$ of G-invariant C^2-functionals defined on the G-Hilbert space X and assume $\nabla\Phi_\lambda(0) = 0$ for all $\lambda \in \mathbb{R}$. Suppose moreover that for some $\lambda_0 \in \mathbb{R}$ the Hessian of Φ_{λ_0} at 0 has a non-trivial kernel $Y \subset X$. Under what conditions on Φ do non-trivial solutions of $(*)$ bifurcate from $(\lambda_0, 0)$? What can we say about the number of bifurcating G-orbits of solutions of $(*)$?

We shall deal with these and related problems in chapter 7. If one is only interested in equation $(*)$ and if one takes $\|x\|$ as a parameter for the bifurcating solutions then Problem 2.23 can be reformulated in order to fit under the general setting of Theorem 2.19. In that case one only has to study the existence of critical points of $\Phi : SY \rightarrow \mathbb{R}$ where Y is the eigenspace of an eigenvalue λ_0 of the Hessian of Φ at 0 and SY is a small sphere in Y around 0. But if the equation has the more general form $\nabla\Phi_\lambda(x) = 0$ or if one wants to parametrize the bifurcating solutions using the natural parameter λ then one needs a different approach. In fact, it does not seem possible to work only with the category or the genus.

We shall see that for both problems 2.22 and 2.23 a cohomological approach to critical point theory is very useful. Concerning the computation of the $\mathcal{A}$-category we merely need a computable lower bound. This will be an equivariant version of the cup-length. It turns out that this "length" has a number of properties which can be used successfully in the study of the bifurcation problem. Some of these properties do not hold for the $\mathcal{A}$-category nor for the $\mathcal{A}$-genus.

2.6 A symmetric mountain pass theorem

In this section we shall prove a generalization of the symmetric mountain pass theorem of Ambrosetti and Rabinowitz [AmR]. We shall also sketch an application of our result to a non-linear elliptic system with Dirichlet boundary conditions. This serves to illustrate two points. First we shall not just quote Theorem 2.19 but instead work directly with the genus. This approach seems to be easier and it allows us to demonstrate the flexibility of the genus. Second, although Theorem 2.19 already indicates the importance of the proper choice of $\mathcal{A}$ we shall work here with a completely different set $\mathcal{A}$. In the end we shall mention how one can use (a modified version of) Theorem 2.19 to prove the mountain pass theorem. Before we can state it we need the following definition.

2.24 Definition:

Let V be a finite-dimensional representation space of G. V is said to be *admissible* if every G-map $\overline{\mathcal{O}} \rightarrow V^{k-1}$, $\mathcal{O}$ an open, bounded, invariant neighborhood of 0 in V^k, $k \geq 1$, has a zero in $\partial\mathcal{O}$. In other words, there does not exist a G-map $\partial\mathcal{O} \rightarrow V^{k-1} - 0$. Here G acts diagonally on V^l.

It is clear that an admissible representation V must have a trivial fixed point set: $V^G = 0$. It is also clear that a representation of $G = \mathbb{Z}/2$ with trivial fixed point set is admissible by the classical Borsuk-Ulam theorem. We can now formulate the mountain pass theorem.

2.25 Theorem:

Let V be an admissible representation of G. Let $E = \bigoplus_{j=1}^{\infty} E^j$ be a G-Hilbert space such that each summand E^j is isomorphic to V as a representation of G. Consider a G-invariant functional $\Phi \colon E \to \mathbb{R}$ which satisfies the following hypotheses.

(Φ_1) The Palais-Smale condition holds above $\Phi(0)$, i.e. any sequence u_j in E with $\Phi(u_j) \to c > \Phi(0)$ and $\|\nabla\Phi(u_j)\| \to 0$ has a convergent subsequence.

(Φ_2) For each $k \geq 1$ there exists $R_k > 0$ such that $\Phi(u) \leq \Phi(0)$ for every $u \in E_k = \bigoplus_{j=1}^{k} E^j$ with $\|u\| \geq R_k$.

(Φ_3) There exist $k_0 \geq 1$ and $b > \Phi(0)$, $\rho > 0$ such that $\Phi(u) \geq b$ for every $u \in E_{k_0}^{\perp}$ with $\|u\| = \rho$.

Then Φ possesses an unbounded sequence of critical values $c_k \to \infty$. In fact, these can be characterized as $c_k = \inf_{h} \sup \Phi\big(h(B_k)\big)$ where h runs through all G-maps of the form $h \colon B_k = \{u \in E_k \colon \|u\| \leq R_k\} \to E$ with $h(u) = u$ if $\|u\| = R_k$.

Theorem 2.25 generalizes the mountain pass theorem of Ambrosetti and Rabinowitz who considered even functionals $\Phi \colon E \to \mathbb{R}$. In this case $V = \mathbb{R}$ with the antipodal action of $G = \mathbb{Z}/2$. The classical Borsuk-Ulam theorem tells us that this V is admissible. The mountain pass theorem of [BCP] allows the summands E^j to be different representations of G, depending on j. On the other hand, in [BCP] not all admissible representations are allowed. For example, if G is finite it is assumed that a p-Sylow subgroup G_p of G acts on $E_{k_0}^{\perp}$ without fixed points. We shall see in chapter 3 that V is admissible if $V^{G_p} = 0$ but that the class of admissible representations is much larger.

Proposition 2.20 indicates that for tori and p-tori every representation V with $V^G = 0$ should be admissible. We shall see later that this is true. In fact, it will be shown that V is admissible if there does not exist a G-map $SV^k \to SV^{k-1}$. Thus it would suffice in Definition 2.24 to restrict $\mathcal{O}$ to be the unit ball in V^k. It is also clear that $\mathcal{A}_{SE}$-cat$(SE) = \infty$ for any G-Hilbert space E as in 2.25. So every G-invariant C^1-functional $\Phi \colon SE \to \mathbb{R}$ which satisfies the Palais-Smale condition has an unbounded sequence of critical values. The usefulness of Theorem 2.25 depends of course on how big the class of admissible representations is. This leads to the following questions.

2.26 Problems:

Is it true that every representation V with $V^G = 0$ is admissible? If not, can one characterize those compact Lie groups for which this is the case? Even more, can one characterize admissible representations?

We shall deal with this problem in chapter 3, the relevant result is Theorem 3.7; see also Example 3.20. For the moment we come back to the mountain pass theorem.

Proof of the mountain pass theorem 2.25:

As mentioned in the beginning of this section our main tool will be the $\mathcal{A}$-genus. Our choice of $\mathcal{A}$ is $\mathcal{A} := \{SV\}$. This will be fixed throughout this section. The following result is obvious. It obtains its strength only together with Theorem 3.7.

2.27 Proposition:

If V is admissible and $\mathcal{O} \subset V^k$ is an open, bounded and invariant neighborhood of 0 then $\mathcal{A}$-genus$(\partial\mathcal{O}) = k$.

The proof of the mountain pass theorem parallels the one for the $\mathbb{Z}/2$-version in [Ra2], pp. 55-59; cf. also [Ba5]. Therefore we only give a sketch-proof emphasizing where the admissibility of V enters. It is clear that the c_k are critical values. Unfortunately we are not able to show directly that $c_k \to \infty$ as $k \to \infty$. Instead we define another sequence c'_k of critical values such that $c'_k \leq c_k$ and $c'_k \to \infty$. Let Γ_k be the set of all compact invariant subsets B of E of the form $B = h(\overline{B_m - Y})$, where $m \geq k$, $h: B_m \to E$ is a G-map with $h(u) = u$ if $\|u\| = R_m$ and Y is a closed invariant subset of B_m with $\mathcal{A}$-genus$(Y) \leq m - k$. For $k \in \mathbb{N}$ we set

$$c'_k := \inf_{B \in \Gamma_k} \sup \Phi(B).$$

Since $h(B_k) \in \Gamma_k$ we have $c_k = \inf_h \sup \Phi(h(B_k)) \geq c'_k$.

2.28 Lemma:

If $k > k_0$ then $c_k \geq b > 0$.

Proof:

We show that for every $B \in \Gamma_k$

$$B \cap E_{k_0}^\perp \cap \{\|u\| = \rho\} \neq \emptyset.$$

Then the lemma follows from (Φ_4). Consider an element $B = h(\overline{B_m - Y})$ of Γ_k and set $\mathcal{O} = \{u \in B_m: \|h(u)\| < \rho\}$. Since $h(0) = 0$ (h is equivariant and $V^G = 0$) and $h(u) = u$ for $u \in \partial B_m$ the set $\mathcal{O}$ is a bounded, open, invariant neighborhood of 0 in E_m contained in the interior of B_m. For $u \in \partial\mathcal{O}$ we must have $\|h(u)\| = \rho$. Setting $X = \overline{\partial\mathcal{O} - Y} \subset \partial\mathcal{O}$ we find $h(X) \subset \{u: \|u\| = \rho\}$. Using the continuity of the genus, the subadditivity property also holds with closed subsets instead of open subsets. This and Proposition 2.27 yield

$$\mathcal{A}\text{-genus}(X) \geq \mathcal{A}\text{-genus}(\partial\mathcal{O}) - \mathcal{A}\text{-genus}(Y) \geq m - (m - k) = k.$$

Now we apply the monotonicity of the $\mathcal{A}$-genus and once more Proposition 2.27 to obtain

$$\mathcal{A}\text{-genus}(\{u \in E - E_{k_0}^\perp: \|u\| = \rho\}) = \mathcal{A}\text{-genus}(\{u \in E_{k_0}: \|u\| = \rho\}) = k_0.$$

Thus there cannot exist a G-map $X \to \{u \in E - E_{k_0}^\perp: \|u\| = \rho\}$ because of the assumption $k > k_0$. This implies $B \cap \{\|u\| = \rho\} \cap E_{k_0}^\perp \supset h(X) \cap E_{k_0}^\perp \neq \emptyset$, as required. □

2.29 Lemma:

   If $k > k_0$ and $c'_k = \ldots = c'_{k+p} = c$ then $\mathcal{A}$-genus$(K_c) \geq p + 1$ where K_c is the set of critical points of Φ with critical value c.

Proof:

   The proof goes as in [Ra2], Proposition 9.30.   □

2.30 Lemma:

   $c'_k \to \infty$ *as* $k \to \infty$.

Proof:

   Obviously $c'_k \leq c'_{k+1}$. If $c'_k \to c < \infty$ then K_c is compact by the Palais-Smale condition (Φ_1). Here we used Lemma 2.28. Moreover, $0 \notin K_c$ since $\Phi(0) < b \leq c$. Proposition 2.16 yields $\mathcal{A}$-genus$(K_c) < \infty$, hence $c_k < c$ for all k by Lemma 2.29. Now one can proceed as in the proof of Proposition 9.31 in [Ra2].   □

   It is also possible to reduce the mountain pass theorem to Theorem 2.19. One can show that $\mathcal{A}$-cat$_E(E, \Phi^0) = \infty$ with $\mathcal{A} = \{SV\}$. It is not difficult to see that this suffices to obtain an unbounded sequence of critical values of Φ because every nontrivial G-orbit in E can be deformed into some $SV \subset E$. To compute the $\mathcal{A}$-category one can estimate it from below by a relative version of the $\mathcal{A}$-genus similar to the one defined in Remark 2.8b) of [BC2]. Also the admissibility condition has to be replaced by a relative formulation similar to Theorem A.1 of [BC2]. Finally, knowing all admissible representations (see Theorem 3.7 in the next chapter) one checks that they also satisfy the relative admissibility condition. This proof is not any more difficult then the one given here. But for us it was certainly easier to refer to the proof of the mountain pass theorem in [Ra2].

   As an application of the mountain pass theorem we study the non-linear Dirichlet problem

$$(D) \qquad \begin{array}{ll} \Delta u + F_u(u) = 0 & \text{in } \Omega \subset \mathbb{R}^n, \\ u = 0 & \text{on } \partial\Omega. \end{array}$$

Here Ω is an open and bounded subset of $\mathbb{R}^n$ with smooth boundary. We assume the following hypotheses on F.

(F_1) $F: \mathbb{R}^m \to \mathbb{R}$ is C^1.

(F_2) There are constants $C \geq 0$ and $1 < s < (n+2)/(n-2)$ if $n > 2$ such that

$$|F_u(u)| \leq C \cdot (1 + |u|^s).$$

If $n = 2$ this assumption can be relaxed, if $n = 1$ it can be omitted.

(F_3) There are constants $R > 0$ and $\mu > 2$ such that $0 < \mu F(u) \leq u \cdot F_u(u)$ if $|u| \geq R$.

(F_4) There exists a compact Lie group G acting orthogonally on $V = \mathbb{R}^m$ such that V is an admissible representation of G and $F: V \to \mathbb{R}$ is invariant under this action.

2.31 Theorem:

If F sasifies the hypotheses $(F_1) - (F_4)$ then (D) possesses an unbounded sequence of weak solutions in $W_0^{1,2}(\Omega, \mathbb{R}^m)$. These solutions are classical solutions of (D) if moreover F is $C^{1+\alpha}$, $0 < \alpha < 1$, in every compact subset of $\mathbb{R}^m$.

If $G = \mathbb{Z}/2$, $m = 1$ and F is even, Theorem 2.31 is due to Ambrosetti and Rabinowitz [AmR]. And if G is a cyclic group acting freely on V then the result is due to Michalek [Mi]. Of course, if there exists a cyclic subgroup H of G such that V is a free H-space then one may again quote [Mi] in order to obtain the conclusion of Theorem 2.31. But we shall see that the class of admissible representations is much bigger.

We could also consider more general situations. For instance, $-\Delta$ could be replaced by a second order uniformly elliptic operator with divergence structure. Also F could depend on $x \in \overline{\Omega}$, that is $F \in C^1(\overline{\Omega} \times \mathbb{R}^m, \mathbb{R})$. But we avoided these technicalities since the main goal of this section is to illustrate the use of the genus and to motivate the study of admissible representations.

Sketch-Proof of Theorem 2.31:

Set $E = W_0^{1,2}(\Omega, \mathbb{R}^m)$ and consider the functionals

$$\Psi \colon E \to \mathbb{R}, \ \Psi(u) = \int_\Omega F(u)\, dx$$

and

$$\Phi \colon E \to \mathbb{R}, \ \Phi(u) = \frac{1}{2} \int_\Omega |\nabla u|^2\, dx - \Psi(u).$$

By the Poincaré inequality we may take $\|u\| = (\int_\Omega |\nabla u|^2\, dx)^{1/2}$ as a norm in E. So $\Phi(u) = \frac{1}{2}\|u\|^2 - \Psi(u)$. As a consequence of (F_1) and (F_2) Φ is in $C^1(E, \mathbb{R})$; see [Ra2], Appendix B. A critical point $u \in E$ of Φ satisfies

$$D\Phi(u)v = \int_\Omega (\nabla u \cdot \nabla v - F_u(u) \cdot v)\, dx = 0$$

for all $v \in E$ and is by definition a weak solution of (D). Standard regularity theory shows that u is in fact a classical solution provided F is $C^{1+\alpha}$, $0 < \alpha < 1$, on compact subsets of $\mathbb{R}^m$. Setting $v = u$ in the above equation we obtain

$$\|u\|^2 = \int_\Omega F_u(u) \cdot u\, dx$$

and

$$\Phi(u) = \int_\Omega \left(\frac{1}{2} F_u(u) \cdot u - F(u) \right) dx.$$

Thus if we can find a sequence u_k of critical points of Φ such that $\Phi(u_k) \to \infty$ as $k \to \infty$ then these two equations together with hypothesis (F_3) imply $\|u_k\| \to \infty$.

The action of G on $V = \mathbb{R}^m$ induces an action of G on E:

$$(gu)(x) := g\big(u(x)\big) \quad \text{for every } g \in G, \ u \in E \text{ and } x \in \Omega.$$

This turns E into a G-Hilbert space, i.e. each $g \in G$ acts on E as an orthogonal linear map. Moreover, Ψ and Φ are invariant because F is. Next let $\lambda_1 < \lambda_2 \leq \lambda_3 \leq \ldots, \lambda_j \to \infty$, be the eigenvalues of the problem

$$-\Delta v = \lambda v \qquad \text{in } \Omega,$$
$$v = 0 \qquad \text{on } \partial\Omega,$$

and $v^1, v^2, v^3, \ldots$ the corresponding eigenfunctions. The v^j form a complete set of orthogonal eigenvectors of $W_0^{1,2}(\Omega)$. For $j \geq 1$ set

$$E^j = \{u = (u_1, \ldots, u_m) \in E : u_i = \alpha_i v^j, \alpha_i \in \mathbb{R} \text{ for } i = 1, \ldots, m\}$$

and

$$E_k = \bigoplus_{j=1}^{k} E^j \subset E.$$

Observe that each E^j is a G-invariant linear subspace of E isomorphic to V as a representation space of G. E is spanned by the E^j. Theorem 2.31 is a consequence of the mountain pass theorem 2.25. One only has to check the hypotheses $(\Phi_1) - (\Phi_3)$. This can be done as in the proof of Theorem 9.38 in [Ra2] where the case $G = \mathbb{Z}/2$ and $V = \mathbb{R}$ has been treated; see also [Ba5]. $\qquad\qquad \square$

Chapter 3

Category and genus
of infinite-dimensional representation spheres

3.1 Introduction

In this chapter we address the problems posed in 2.22 and 2.26.

1. Is the category or the genus of an infinite-dimensional sphere in a G-Hilbert space without fixed points necessarily infinite?

2. Is a representation V without non-trivial fixed points admissible? In particular, can one exclude the existence of G-maps $SV^k \to SV^{k-1}$ between the unit spheres of V^k and V^{k-1}?

We know already that both questions can be answered in the affirmative if G is a p-torus $(\mathbb{Z}/p)^k$ or a torus $(S^1)^k$. The same is obviously true for arbitrary G if a subgroup H of G which is isomorphic to a $(p$-$)$torus acts without fixed points on the sphere, for example if the action of G on the sphere is free. Here we shall classify those groups for which all infinite-dimensional fixed point free representation spheres have infinite category (respectively genus). And we shall classify all admissible representations. In order to prove that the category is infinite we need an equivariant cup-length estimate. This part of the proof is deferred to §5.4. The basic idea to find counterexamples is to construct G-maps $SV \to SW$ where W is a proper subrepresentation of V and $W^G = 0$. The existence of such a map can be considered as an anti-Borsuk-Ulam theorem. The main results of this chapter are stated in §3.2. In §3.3 we prove our anti-Borsuk-Ulam theorems and in §3.4 we use these to construct infinite-dimensional representation spheres with finite category. Finally, in §3.5 we illustrate these results by looking at representations of the cyclic groups $\mathbb{Z}/pq$ with two different primes p and q, and at representations of $SO(3)$. We also discuss the existence of Borsuk-Ulam type theorems for p-groups and p-toral groups. In particular, we give estimates for the maximal real number a_G with the property that the existence of a G-map $SV \to SW$ implies $\dim W \geq a_G \cdot \dim V$.

The methods used in sections 3.3 and 3.4 center around equivariant obstruction theory. They are not needed elsewhere in the book. Section 3.5 can be read independently of §§3.3 and 3.4. In this chapter by a representation of G we shall always mean a finite-dimensional representation. If we allow infinitely many dimensions we shall talk about G-Hilbert spaces or more generally about normed G-vector spaces. We recall that these are possibly infinite-dimensional vector spaces (over $\mathbb{R}$) with a linear action of G preserving the norm. A normed G-vector space V is said to be fixed point free if $V^G = 0$.

3.2 Statement of results

Before computing any $\mathcal{A}$-cat(X) we have to fix a set $\mathcal{A}$ of G-ANRs. A reasonable choice is $\mathcal{A} = \mathcal{A}_X = \{G/G_x : x \in X\}$. This excludes the trivial case where already a single G-orbit of X has infinite category because it cannot be mapped into an element of $\mathcal{A}$. And it is the most important case for applications where one wants to count (critical) G-orbits; cf. Theorem 2.19.

3.1 Theorem:

a) *Let G be a p-toral group, i.e. an extension of a finite p-group P by a torus T:*
$1 \to T \to G \to P \to 1$. *Then for any (non-equivariantly) contractible, fixed point free G-ANR X: $\mathcal{A}_X$-cat$(X) = \mathcal{A}_X$-genus$(X) = \infty$. This applies, for instance, to spheres of infinite-dimensional, fixed point free normed G-vector spaces.*

b) *If G is not p-toral then there exists an infinite-dimensional fixed point free G-Hilbert space V with $\mathcal{A}_{SV}$-genus$(SV) \le \mathcal{A}_{SV}$-cat$(SV) < \infty$.*

If G is not p-toral one needs a criterion when $\mathcal{A}_{SV}$-cat$(SV) = \infty$ for a given G-Hilbert space V. More generally, let V be a normed G-vector space.

3.2 Corollary:

$\mathcal{A}_{SV}$-cat$(SV) = \mathcal{A}_{SV}$-genus$(SV) = \infty$ *if there exist closed subgroups $K \subset H$ of G with K normal in H and H/K a p-toral group such that $V^H = 0$ and $\dim V^K = \infty$.*

We conjecture that $\mathcal{A}_{SV}$-genus$(SV) < \infty$ if subgroups H, K as in 3.2 do not exist. Concerning the category we are not so optimistic that the corollary provides a complete classification of those V with $\mathcal{A}_{SV}$-cat$(SV) = \infty$. But for a given group G we expect only a small number of counterexamples V where $\mathcal{A}_{SV}$-cat$(SV) = \infty$ although the corollary cannot be applied (cf. Examples 3.19 and 3.20).

Corollary 3.2 combines two *principles of symmetry reduction* which are elementary but (or, hence) used quite frequently. If we set $K = \{e\}$ the trivial group then we can consider $V = V^K$ as an H-space for a subgroup H of G. Of course, H should be easier to deal with than G. In our case, $H = H/\{e\}$ is p-toral so that Theorem 3.1 applies. But one has to be cautious when interpreting the results obtained by forgetting part of the group. Observe that $\mathcal{A}_{SV}$-cat(SV) depends heavily upon whether SV is considered as a G-space or as an H-space, and in the end our result should hold for the G-space. The second way to reduce the symmetry from a big group G to a nicer group consists of passing from V to a fixed point space V^K. This is only an NK-space where NK is the normalizer of K in G. Since K acts trivially on V^K we get a representation of $WK := NK/K$. If WK is "nice", that is p-toral in our situation, set $H = NK$ in Corollary 3.2 and apply Theorem 3.1 to the WK-space V^K. We shall encounter these principles of symmetry reduction in our applications in chapters 8 and 9. There $G = O(3)$ or $G = S^1 \rtimes \Gamma$, a semidirect product of S^1 and an arbitrary Lie group Γ. These groups are certainly not "nice" from our point of view.

Another variation of this type of result is the following one.

3.3 Corollary:
Let X be a contractible fixed point free G-ANR. Then

$$\mathcal{A}_X\text{-cat}(X) = \mathcal{A}_X\text{-genus}(X) = \infty$$

if the following equivalent conditions hold.

(i) *There exists a prime p which divides the Euler characteristic $\chi(Gx) = \chi(G/G_x)$ of any orbit of X.*

(ii) *There exists a prime p such that a p-toral Sylow subgroup G_p of G acts on X without fixed points.*

Here G_p is defined as follows. Let T be a maximal torus of G, NT its normalizer and $\pi\colon NT \to W := NT/T$ the projection onto the Weyl group. W is always finite. Then $G_p := \pi^{-1}(W_p)$ where W_p is a p-Sylow subgroup of W.

Corollaries 3.2 and 3.3 contain in particular the case of free actions, an assumption made frequently although it is very special. The only compact Lie groups which can act freely (and linearly) on representation spheres are S^1, $S^3 = SU(2)$, the normalizer of S^1 in S^3 and certain finite groups listed in [Wo], Theorems 6.1.11 and 6.3.1; see [Bre], Theorem III.8.5.

At first sight it is surprising that $\mathcal{A}_{SV}\text{-cat}(SV)$ can be finite for an infinite-dimensional fixed point free normed G-vector space V. After all, this implies that SV/G has finite Lusternik-Schnirelmann category. And isn't SV/G something like an infinite projective space $\mathbb{F}P^\infty$ or even more complicated? But for certain groups the situation can be even more strange.

3.4 Theorem:
Suppose G is either a finite group containing no normal subgroup of prime index (for example a simple group) or a connected semisimple Lie group (i.e. the center of the group is finite). Then there exists a (finite-dimensional) fixed point free representation W such that the following holds:

a) *If $V = W \oplus U$ where U is any normed G-vector space with $U^G = 0$ then*

$$\mathcal{A}_{SV}\text{-cat}(SV) \le 2 \cdot \mathcal{A}_{SW}\text{-cat}(SW) < \infty$$

independently of U.

b) *If X is any fixed point free G-ANR with $\mathcal{A}_{SW} \subset \mathcal{A}_X$, for example $X = SV$ as in a), then*

$$\mathcal{A}_X\text{-genus}(X) \le \mathcal{A}_{SW}\text{-genus}(SW) < \infty.$$

As a consequence of 3.4b), setting $\mathcal{A} = \{G/H \colon H \subsetneq G$ a closed subgroup$\}$ then

$$\mathcal{A}\text{-genus}(X) \leq \mathcal{A}_{SW}\text{-genus}(SW)$$

for any fixed point free G-ANR X. Thus $\mathcal{A}$-genus is bounded! The situation for $\mathcal{A}$-cat is not much better. From the critical point theory point of view the results obtained so far tell us that the $\mathcal{A}$-category and the $\mathcal{A}$-genus are not very helpful once one leaves the realm of p-groups — except if additional assumptions like those of Corollary 3.2 are satisfied.

Closely related to the computation of $\mathcal{A}_{SV}$-cat(SV) and $\mathcal{A}_{SV}$-genus(SV) is the following theorem about the existence or non-existence of G-maps between representation spheres. Results of this kind are often called Borsuk-Ulam type theorems.

3.5 Theorem:
   For a compact Lie group G the following are equivalent:
   (i) *If there exists a G-map $SV \to SW$ between spheres of G-Hilbert spaces with $\dim V = \infty$ and $W^G = 0$ then $\dim W = \infty$.*
  (ii) *If there exists a G-map $SV \to SW$ between representation spheres with $W \subset V$ and $W^G = 0$ then $W = V$.*
 (iii) *G is p-toral.*

Theorem 3.4 is a consequence of

3.6 Theorem:
   Suppose that either G is a finite group containing no normal subgroup of prime index or a connected semisimple Lie group. Then there exists a fixed point free representation W such that the following holds: For any fixed point free G-ANR X there exists a G-map $X \to SW$.

The representations W appearing in 3.4 and 3.6 are by no means unique. As the proofs in §§3.3, 3.4 show the possible choices of W for a given group G can be determined explicitly. It is clear that $W^H \neq 0$ for all proper closed subgroups H of G because otherwise there is no G-map $G/H \to SW$. If W is a complex representation satisfying $\dim_{\mathbb{C}} W^H \geq 2$ for all $H \subsetneq G$ then there exists a G-map $X \to SW$ for any fixed point free G-ANR X. If X is finite-dimensional then it suffices that $\dim_{\mathbb{C}} W^H \geq 1$ for every $H \subsetneq G$. In Example 3.20 in §3.5 below we shall determine all possible W for $SO(3)$.

We conclude this section by classifying all admissible representations. Recall from Definition 2.24 that a representation V of G is admissible if every G-map $\partial \mathcal{O} \to V^{k-1}$ has a zero. Here $\mathcal{O}$ is an open bounded invariant neighborhood of 0 in V^k. If V is admissible and X is a normed G-vector space containing V^n then $\mathcal{A}_{SV}$-genus$(SX) \geq \{SV\}$-genus$(SX) \geq n$. For applications to critical point theory this estimate is only useful if all critical G-orbits in which one is interested are G-homeomorphic to orbits of SV, in particular if $\mathcal{A}_{SV} = \mathcal{A}_{SX}$. This is the case in §2.6 where the G-Hilbert space E is the orthogonal sum of the subspaces E^j, $j \in \mathbb{N}$, and each E^j is isomorphic to V. Here we get $\mathcal{A}_{SE}$-genus$(SE) = \mathcal{A}_{SV}$-genus$(SE) = \infty$. We invite the reader to guess for which groups all fixed point free representations are admissible. Here is the answer.

3.7 Theorem:

a) *For a compact Lie group G the following are equivalent:*

(i) *Every fixed point free representation is admissible.*

(ii) *G is solvable, i.e. the connected component G_0 of G is a torus and the factor group G/G_0 is solvable.*

b) *A representation V of G is admissible iff there exist subgroups $K \subset H$ of G with K normal in H and H/K a solvable group such that $V^H = 0$ and $V^K \neq 0$.*

Thus if the concept of admissibility suffices for an application (as is the case of the mountain pass theorem 2.25) then one can deal with more general groups and more general representations than those allowed by $\mathcal{A}$-cat or $\mathcal{A}$-genus. More precisely, if one has to exclude maps $SV \to SW$ with $W \subsetneq V$ and $W^G = 0$ then one better finds a reduction from G to a p-toral group. But if one only has to exclude maps $SV^k \to SV^{k-1}$ then it suffices to find a reduction to a solvable group. So it is important to analyze precisely which kind of Borsuk-Ulam theorem is needed.

3.3 G-maps between spheres

The goal of this section is the construction of G-maps between representation spheres. We start with the following result which is a part of Theorem 3.5. Its proof will take up all of this section.

3.8 Proposition:

If G is not p-toral then there exist representations $W \subsetneq V$ with $W^G = 0$ and a G-map $SV \to SW$.

Proof:

The basic idea for the construction of such maps is best illustrated in the case $G = \mathbb{Z}/pq$ where p and q are different primes; cf. also [Wa2]. Let G act on $W \cong \mathbb{C}^2$ via $g(z_1, z_2) := (g^p z_1, g^q z_2)$ where we consider G as a subgroup of $S^1 \subset \mathbb{C}$. Conner and Floyd constructed a G-map $f \colon SW \to SW$ of degree zero; see [ConnF], (3.1), or the book by Bredon [Bre], I.8(B). Next let G act on $U \cong \mathbb{C}$ via scalar multiplication and set $V = W \oplus U$. Then (SV, SW) is a G-CW-pair and SV is obtained from SW by attaching G-cells of the form $G \times e^n$, $0 \leq n \leq 5$, because G acts freely on $SV - SW$. We write X_n for the relative n-skeleton; thus $X_{-1} = SW$ and $X_5 = SV$. Suppose we have a G-map $f_{n-1} \colon X_{n-1} \to SW$ for some $n \leq 5$. We claim that we can extend $f \circ f_{n-1}$ to X_n. Let $\varphi \colon G \times \partial e^n \to X_{n-1}$ be an attaching map. The composition $f \circ f_{n-1} \circ \varphi \colon \{e\} \times \partial e^n \to SW$ is nullhomotopic because $\deg(f) = 0$. Therefore we can extend this composition to a (non-equivariant) map $\{e\} \times e^n \to SW$. This can be extended further to a G-map $G \times e^n \to SW$ in the obvious way. Of course, if $n \leq \dim SW$ then we can extend any G-map $f_{n-1} \colon X_{n-1} \to SW$ to X_n and need not consider $f \circ f_{n-1}$.

Thus if $G = \mathbb{Z}/pq$ there exists a G-map $SV \to SW$ as in 3.8. If G is finite and contains an element g of mixed order pq set $H := \langle g \rangle \cong \mathbb{Z}/pq$. Let $W \subsetneq V$ be representations of H and $f: SV \to SW$ be an H-map as in 3.8. Upon passing to the induced representations we obtain a G-map

$$\mathrm{ind}_H^G f: S(\mathrm{ind}_H^G V) \to S(\mathrm{ind}_H^G W)$$

as claimed. Here

$$\mathrm{ind}_H^G V = \{\alpha: G \to V \mid \alpha(gh) = h^{-1} \cdot \alpha(g) \text{ for } g \in G \text{ and } h \in H\}$$

is a representation of G via $(g\alpha)(g') := \alpha(g^{-1}g')$. The norm on $\mathrm{ind}_H^G V$ is defined as $\|\alpha\| = (\sum_{g \in G} \|\alpha(g)\|^2)^{1/2}$ and the map $F := \mathrm{ind}_H^G f$ is given by

$$(F(\alpha))(g) = \widetilde{f}(\alpha(g)) \in W \quad \text{where } \widetilde{f}: V \to W \text{ is the radial extension of } f.$$

Since $\widetilde{f}$ preserves the norm so does F.

This proves Proposition 3.8 for all finite groups which contain an element of mixed order. The idea in the general case will be the same. We start with a G-map $SW \to SW$ and try to extend it to a G-map $S(W \oplus U) \to SW$ for certain representations W, U of G.

3.9 Lemma:

Let W be a representation of G and $f: SW \to SW$ be a G-map. Let (X, Y) be a finite-dimensional relative G-CW-complex such that $\deg(f^H) = 0$ for every relative G-cell $G/H \times e^n$. We use the convention $\deg(f^H) = 1$ if $W^H = 0$. Then there exists a G-map $X \to SW$ iff there exists a G-map $Y \to SW$.

Proof:

Let $f: Y \to SW$ be a G-map and write X_n for the relative n-skeleton of (X, Y) so that $X_{-1} = Y$. We first extend f over the 0-skeleton as follows. If $G/H \times e^0$ is a 0-cell of $X - Y$ then $\deg(f^H) = 0$, hence $SW^H \neq \emptyset$. Choose any $w \in SW^H$ and set $f_0(gH \times e^0) := gw$. This defines a G-map $f_0: X_0 \to SW$. Suppose we have a G-map $f_{n-1}: X_{n-1} \to SW$ for some $n \geq 1$. Then we can extend $f \circ f_{n-1}$ to X_n as follows. If $\varphi: \coprod_\alpha G/H_\alpha \times \partial e^n \to X_{n-1}$ is the attaching map for the n-cells then the restriction $f \circ f_{n-1} \circ \varphi | \coprod_\alpha \{eH_\alpha\} \times \partial e^n$ maps each summand $\{eH_\alpha\} \times \partial e^n$ to SW^{H_α} and is nullhomotopic because $\deg(f^{H_\alpha}) = 0$ for any possible α. Therefore we can extend this restriction to a map $\coprod_\alpha \{eH_\alpha\} \times e^n \to SW$ where each summand is again mapped (non-equivariantly) into the appropriate fixed point space. Finally, this map can be extended uniquely to a G-map $\coprod_\alpha G/H_\alpha \times e^n \to SW$, hence we obtain a G-map $X_n \to SW$. $\square$

Now we are interested in the possible degrees $\deg(f^H)$ for G-maps $f: SW \to SW$ and subgroups H of G. Observe that $\deg(f^H) = \deg(f^K)$ if H and K are conjugate so $\deg(f^H)$ depends only on the conjugacy class (H) of H. Let $\psi(G)$ be the space of conjugacy classes of closed subgroups of G. It is a totally disconnected compact Hausdorff space partially ordered by subconjugation (see [Di], Section IV.3). A G-map $f: SW \to SW$ induces a continuous map

$$\psi(G) \to \mathbb{Z}, \quad (H) \mapsto \deg(f^H),$$

the *degree function* of f. We write $C(G)$ for the set of continuous functions $\psi(G) \to \mathbb{Z}$.

3.10 Proposition:

Let V be a complex representation of the compact Lie group G. Then a function $d \in C(G)$ is the degree function of a G-map $SV \to SV$ if and only if for each pair of subgroups K, H of G the following conditions hold:

(i) If K is a normal subgroup of H and H/K is a torus then $d(H) = d(K)$.

(ii) If H has finite index in its normalizer ($WH = NH/H$ is finite) then the following congruence holds:

$$\sum_{(L)} m(H, L) \cdot d(L) \equiv 0 \quad (\text{mod } |WH|).$$

Here (L) runs through all NH-conjugacy classes of subgroups $H \subset L \subset NH$ such that L/H is finite cyclic. The integer $m(H, L)$ is equal to $|NH/NH \cap NL| \cdot |(L/H)^*|$ where $(L/H)^*$ is the set of generators of the cyclic group L/H.

(iii) If $\dim W^H = 0$ then $d(H) = 1$.

(iv) If $W^H = W^K$ then $d(H) = d(K)$.

Proof:

The proposition is an immediate consequence of Theorems II(5.17) and IV(5.8) of [Di].   □

Now we consider the case G finite. Let $V_0, \ldots, V_r$ be a complete set of irreducible complex representations of G with V_0 denoting the trivial representation and set $W_0 := V_1 \oplus \ldots \oplus V_r$.

3.11 Corollary:

Consider a finite group G and the representation W_0 as above. A function $d \in C(G)$ is the degree function of some G-map $SW_0 \to SW_0$ iff $d(G) = 1$ and for each $(H) \in \psi(G)$ the congruence 3.10(ii) holds.   □

Proof:

It only remains to check the assumptions (iii) and (iv) from Proposition 3.10. These are clear because by our choice of W_0 we have $W_0^H = 0$ only for $H = G$ and $W_0^H = W_0^K$ only for $H = K$.   □

An immediate consequence of Corollary 3.11 is the following one.

3.12 Corollary:

Suppose G is finite and there does not exist a normal subgroup of prime index. Then there exists a G-map $f: SW_0 \to SW_0$ with $\deg(f^H) = 0$ for all $H \subsetneq G$.   □

Combining this result with Lemma 3.9 we obtain the next corollary.

3.13 Corollary:

   Let G and W_0 be as in 3.12. For any finite-dimensional fixed point free G-CW-complex X there exists a G-map $X \to SW_0$. In particular, for any representation U of G with $U^G = 0$ there exists a G-map $S(W_0 \oplus U) \to SW_0$.   □

   Now we prove Proposition 3.8 for finite groups. It remains to consider a finite group G where all elements have prime power order but which is not a p-group. In order to construct a G-map $SV \to SW$ as required it suffices to find a subgroup H of G and to construct an H-map $f: SV \to SW$ with $W \subsetneq V$, $W^G = 0$. Then we pass to the induced map $\mathrm{ind}_H^G f$. Using this observation and Corollary 3.13 an easy induction argument proves Proposition 3.8 for non-solvable groups. Now assume G is solvable. There are not many classes of groups all of whose elements have prime power order. To treat these cases we recall some group theory. A general reference is the book by Gorenstein [Gor]. The group G is said to be a *Frobenius group* if it is a semidirect product $G = F \cdot K$ of the normal subgroup F by the subgroup K such that $g^{-1}Kg \cap K = \{e\}$ for any $g \in G - K$. F is called the Frobenius kernel and K a Frobenius complement of G. An example of a Frobenius group is the dihedral group of order $2n$ with $F \cong \mathbb{Z}/n$ and $K \cong \mathbb{Z}/2$ or the alternating group A_4 of order 12 with $F \cong \mathbb{Z}/2 \times \mathbb{Z}/2$ and $K \cong \mathbb{Z}/3$.

3.14 Lemma:

   Any solvable group G which is not a p-group either contains an element of mixed order or it contains a Frobenius group $F \cdot K$ with F a p-group and K a q-group where p and q are different primes.

Proof:

   Suppose G contains no element of mixed order. Then the centralizer of any element different from the unit is a p-group, hence nilpotent. Groups in which all centralizers are nilpotent, so-called CN-groups, play an important role in the development of group theory and have been studied extensively. At least one of the following holds (see [Gor], Theorem 14.1.5).

   (i) G is nilpotent.

   (ii) G is a Frobenius group whose Frobenius complement is either cyclic or the direct product of a cyclic group of odd order and a (generalized) quaternion group.

   (iii) G is a *3-step group*, i.e. there exists a normal subgroup N of G which is a Frobenius group $N = F \cdot K$ with F a p-group and K a cyclic group of odd order not divisible by p and G/F is a Frobenius group with kernel $N/F \cong K$ and complement a p-group.

The smallest example of a 3-step group is the symmetric group S_4 where we have $N = A_4 \cong (\mathbb{Z}/2 \times \mathbb{Z}/2) \cdot \mathbb{Z}/3$ and $G/N \cong \mathbb{Z}/2$. Observe that the order of S_4 is 24 but S_4 contains no element of order 6. If G is nilpotent then it is either a p-group or it contains an element of mixed order because nilpotent groups are direct products of their Sylow subgroups ([Gor], Theorem 2.3.5). So (i) cannot hold by our assumptions. If (ii) holds then the Frobenius kernel of G is nilpotent according to a theorem of Thompson ([Gor], Theorem 10.3.1), hence it must be a p-group.

Since the Frobenius complement is cyclic or the direct product of a cyclic group of odd order and a quaternion group it must be a q-group (the quaternion groups are 2-groups). So G is a Frobenius group of the required form. Similarly, if G is a 3-step group it must contain a Frobenius group as above.   □

We come back to the proof of Proposition 3.8 for solvable groups which do not contain an element of mixed order. We want to construct a G-map $SV \to SW$ with $SW \subsetneq SV$ and $SW^G = \emptyset$. According to Lemma 3.14 it suffices to consider the case where G is a Frobenius group $F \cdot K$ with F a p-group and K a q-group. We assume first that F is elementary abelian, that $K \cong \mathbb{Z}/q$ and that the action of K on F is irreducible, i.e. the K-orbit of any non-trivial element of F generates F. Remember that K acts on F by conjugation and the action is free on $F - \{e\}$. We want to apply Lemma 3.9 and Corollary 3.11, so we need a G-map $f\colon SW_0 \to SW_0$ with W_0 as above and a representation U of G such that $\deg(f^H) = 0$ if $\dim U^H > 0$. Let $U_1 \cong \mathbb{C}$ be any non-trivial irreducible representation of F and let $U := \mathrm{ind}_F^G(U_1)$ be the induced representation of G. Then we define a function $d \in C(G)$ of possible degrees $d(H)$ as follows:

$$d(H) = \begin{cases} 0 & \text{if } H \not\subseteq F \text{ or if } H = K, \\ |F| & \text{if } H = F, \\ 1 & \text{if } H = G. \end{cases}$$

By our assumption on G up to conjugacy there are no other subgroups of G (cf. [Fe], (25.3)). We claim that there exists a G-map $f\colon SW_0 \to SW_0$ having this degree function: $\deg(f^H) = d(H)$. We have to check the congruences of Proposition 3.10(ii): For any $(H) \in \psi(G)$

$$\sum_{(L)} m(H, L) \cdot d(L) \equiv 0 \pmod{|NH/H|}.$$

The sum is taken over all NH-conjugacy classes (L) such that H is a normal subgroup of L and L/H is cyclic. The integer $m(H, L)$ is equal to $|NH/NH \cap NL| \cdot |(L/H)^*|$ where $(L/H)^*$ is the set of generators of the cyclic group L/H. If $H = G$ or H is conjugate to K then $NH = H$, hence, the congruence is trivially true. If $H = F$ then

$$\sum_{(L)} m(F, L) \cdot d(L) = m(F, F) \cdot d(F) + m(F, G) \cdot d(G) = |F| + q - 1.$$

Now $K \cong \mathbb{Z}/q$ acts freely on $F - \{e\}$ so that q divides $|F| - 1$. Therefore $|F| + q - 1 \equiv 0$ $(\mathrm{mod}\ q = |NF/F|)$. Next consider a subgroup H of F of index p. Then $NH = F$ is normal in G and

$$\sum_{(L)} m(H, L)d(L) = m(H, H)d(H) + m(H, F)d(F) = (p - 1) \cdot |F|.$$

This is congruent to 0 modulo $p = |NH/H| = |F/H|$. Finally, if H is a subgroup of F of index at least p^2 then the congruence is trivially true since all L with L/F cyclic are proper subgroups of F, so $d(L) = 0$. In conclusion, there exists a G-map $f\colon SW_0 \to SW_0$ as claimed. Lemma 3.9 yields a G-map $S(W_0 \oplus U) \to SW_0$ provided

$U^F = 0$ and $U^G = 0$. It suffices to prove $U^F = 0$. Let $\alpha\colon G \to U_1$ be an element of U^F. This means

$$\alpha(g) = (h\alpha)(g) = \alpha(h^{-1}g)$$

for any $h \in F$ and $g \in G$. This implies that $\alpha|Fg$ is constant for any $g \in G$. Now F is a normal subgroup of G, so α is constant on every gF, hence

$$\alpha(g) = \alpha(gh) = h^{-1} \cdot \alpha(g)$$

for $h \in F$ and $g \in G$. This implies $\alpha(G) \subset U_1^F$. Thus $\alpha = 0$ because $U_1^F = 0$ by assumption.

In conclusion, if $G = F \cdot K$ is a Frobenius group with F elementary abelian, $K \cong \mathbb{Z}/q$ and K acts irreducibly on F then there exists a G-map $SV = S(W \oplus U) \to SW$ as required. We now extend this to the case of an arbitrary Frobenius group $G = F \cdot K$ as in Lemma 3.14. First assume F is elementary abelian and $K \cong \mathbb{Z}/q$. Take any non-trivial element h of F and let P be the subgroup of F generated by the K-orbit of h. Then $H = P \cdot K$ is a Frobenius subgroup of G for which we can construct an H-map $SV \to SW$ as above. Passing to the induced representations we can do this for G, too. Next let F be an arbitrary p-group and $K \cong \mathbb{Z}/q$. Let Φ be the Frattini subgroup of F, i.e. the intersection of all maximal subgroups of F. Then the Frattini factor group $P = F/\Phi$ is elementary abelian according to [Gor], Theorem 5.1.3. Since Φ is a characteristic subgroup of F it is invariant under the action of K. So G/Φ is a Frobenius group with elementary abelian kernel F/Φ and complement $K \cong \mathbb{Z}/q$. Thus there exists a G/Φ-map $SV \to SW$ as required which we can also consider as a G-map. Finally, if $G = F \cdot K$ is an arbitrary Frobenius group as in 3.14 let $h \in K$ be any element of order q and set $Q := \langle h \rangle \subset K$. Then $H = F \cdot Q$ is a Frobenius subgroup of G with complement $Q \cong \mathbb{Z}/q$ for which we can construct an H-map as required and this suffices as usual.

3.15 Remark:

So far we have proved Proposition 3.8 for finite groups. We distinguished between solvable and non-solvable groups. This is not really necessary because Lemma 3.14 also holds for simple groups. According to a theorem of Suzuki (cf. [Su], Theorem 16) there exist precisely 8 simple (non-abelian) groups which do not contain an element of mixed order. These are the projective special linear groups $PSL(3,4)$ and $PSL(2,q)$ for $q = 5, 7, 8, 9, 17$, and the Suzuki groups $Sz(8)$, $Sz(32)$. All of these contain a Frobenius subgroup $F \cdot K$ as in Lemma 3.14 (see for example the books [Hu] and [HuB] for the definitions and some properties of these groups). Thus the construction of equivariant maps $SV \to SW$ for Frobenius groups can also be taken as a starting point for the construction of such maps for non-solvable groups; see [Ba3] for this approach. A simpler proof which does not need as much group theory as the proof given here can be found in [LaiM]. There the authors work with the *Burnside ring* $\Omega(G)$ of G (cf. [Di], p. 19) and apply results of Dress [Dr]. The Burnside ring is somewhat hidden in our argument; see the proof of Lemma 5.12. Still another way (for non-solvable groups) can be found in [Wa2], who also uses [Dr]. There the case $G = \mathbb{Z}/pq$ is also being dealt with.

If G is an arbitrary compact Lie group which is not p-toral then either the connected component G_0 of the identity element is not a torus or the factor group $\Gamma = G/G_0$ is not a p-group. In the latter case Proposition 3.8 is true because any Γ-map $SV \to SW$ can be considered as a G-map. Now assume G_0 is not a torus. It suffices to construct a G_0-map $f\colon SV \to SW$ with $W \subsetneq V$ and $W^G = 0$ because we may then pass to the induced map $\mathrm{ind}_{G_0}^G f$ as usual (G_0 has finite index in G). So we may assume that $G = G_0$ is a compact connected non-abelian Lie group. Moreover, it suffices to consider the case where G is semisimple, i.e. it contains no connected abelian normal subgroup different from the trivial one (equivalently: the center of G is finite). To see this let TZ be the maximal torus of the center of G. Then G/TZ is semisimple as a consequence of the structure theorem of compact connected Lie groups; see [Ho], Theorem XIII.1.3. And any G/TZ-map $SV \to SW$ can be considered as a G-map as usual.

3.16 Lemma:

For a compact connected semisimple Lie group G there exist finitely many closed subgroups $H_1, \ldots, H_r \subsetneq G$ such that any closed subgroup $H \subsetneq G$ is subconjugate to one of the H_i.

Proof:

Recall the space $\psi(G)$ of conjugacy classes of closed subgroups of G. Let $\mu(G)$ be the set of maximal elements of $\psi'(G) := \psi(G) - \{G\}$. Since G is semisimple it is not the limit of proper subgroups of G, that is $\psi'(G)$ is a closed subset of $\psi(G)$; see [Di], Section IV.3 for this and the following arguments. If $(H_i) \in \psi'(G)$ is an increasing sequence then $\lim(H_i) = (H) \in \psi'(G)$ exists and $(H_i) \leq (H)$ for all i. So Zorn's lemma implies that $\mu(G)$ is not empty. Furthermore, $\mu(G)$ must be finite because any sequence (H_i) in $\mu(G)$ contains a convergent subsequence. And if $\lim(H_i) = (H)$ then $(H_i) \leq (H)$ for almost all i, hence $(H_i) = (H)$ for almost all i. $\square$

For G as in Lemma 3.16 there exists a fixed point free complex representation W_0 with $\dim W_0^H \geq 1$ for any $H \subsetneq G$. One simply chooses fixed point free complex representations $V_1, \ldots, V_r$ of G such that there exist elements $v_i \in V_i$ having the isotropy $G_{v_i} = H_i$; see [Di], Theorem I(5.5). Then $W_0 := V_1 \oplus \ldots \oplus V_r$ has the desired property because of Lemma 3.16.

3.17 Lemma:

Let G be as in Lemma 3.16 and let W_0 be a fixed point free complex representation of G such that $\dim W_0^H \geq 1$ for any $H \subsetneq G$. Then there exist G-maps $f\colon SW_0 \to SW_0$ with $\deg(f^H) = 0$ for any $H \subsetneq G$.

Proof:

The function $d: \psi(G) \to \mathbb{Z}$, $d(G) = 1$, $d(H) = 0$ if $H \subsetneq G$, is continuous since $\{G\}$ is a closed and open subset of $\psi(G)$. It is the degree function of some G-map $SW_0 \to SW_0$ if for each $(H), (L) \in \psi(G)$ with H normal in L and L/H a torus $d(H) = d(L)$ and if for any $(H) \in \psi(G)$ with NH/H finite the congruence from Proposition 3.10(ii) is satisfied. In our case these congruences are obvious. Either $H \subsetneq G$ and then $d(L) = 0$ for all possible L since G/L can neither be a torus nor finite (G is connected semisimple). And if $H = G$ then $|NH/H| = 1$. □

Proposition 3.8 is now settled with the next result.

3.18 Corollary:

Let G and W_0 be as in Lemma 3.17 and let X be a finite-dimensional fixed point free G-CW-complex. Then there exists a G-map $X \to SW_0$. In particular, for any (finite-dimensional) representation U of G with $U^G = 0$ there exists a G-map $S(W_0 \oplus U) \to SW_0$.

Proof:

Using Lemma 3.17 instead of Corollary 3.12 the proof is the same as the one of Corollary 3.13. □

3.4 Proofs

There are two types of results in §3.2. One kind states that the category of certain G-spaces is infinite which means that there do not exist finite coverings of these spaces as required in the definition of the category. To prove such a non-existence result some algebraic topology is needed. The same is true for those results where the non-existence of a G-map is claimed. This part of the proof is essentially deferred to §5.4, where we shall prove Theorem 3.1a). In this section we assume Theorem 3.1a) to be true and deduce all other non-existence results.

The second type of results states that the category (or genus) of certain G-spaces is finite or that certain G-maps do exist. Here one has to construct the coverings or G-maps. The results of this kind can all be deduced from the results obtained in §3.3. We begin with the proof of Theorem 3.1b).

Proof of Theorem 3.1b):

If G is not p-toral then Proposition 3.8 yields representations $W \subsetneq V$ with $W^G = 0$ and a G-map $f_1: SV \to SW$. We set $U := W^\perp \subset V$ and define inductively G-maps $f_n: S(W \oplus U^n) \to SW$. Given $n \geq 2$ and a G-map $f_{n-1}: S(W \oplus U^{n-1}) \to SW$ set

$$f_n(w, u_1, \ldots, u_n) := f_1(\tilde{f}_{n-1}(w, u_1, \ldots, u_{n-1}), u_n);$$

here $\tilde{f}_{n-1}: V \to W$ is the radial extension of f_{n-1}. Since $\tilde{f}_{n-1}$ preserves the norm we obtain a G-map $f_n: S(W \oplus U^n) \to SW$ as required. These maps do not commute in general, so they do not induce a G-map $S(W \oplus U^\infty) \to SW$ where we write $S(W \oplus U^\infty)$ for the union $\bigcup_{n=1}^\infty S(W \oplus U^n)$ with the weak topology (induced by the canonical inclusions $S(W \oplus U^n) \hookrightarrow S(W \oplus U^\infty))$. But one can define a G-map $f \colon S(W \oplus U^\infty) \to S(W \oplus W)$ as follows. First replace $S(W \oplus U^\infty)$ by the telescope

$$T = \left(\coprod_{n \geq 1} S(W \oplus U^n) \times [0,1] \right) \Big/ \sim$$

where $(y,1) \in S(W \oplus U^n) \times \{1\}$ is identified with $(y,0) \in S(W \oplus U^{n+1}) \times \{0\}$. Using the equivariant Whitehead theorem 2.5 one sees that T is G-homotopy equivalent to $S(W \oplus U^\infty)$. Now we define $f \colon T \to S(W \times W)$ by setting

$$f(y,t) := \begin{cases} \left(\sqrt{1-t^2} \cdot f_n(y), t \cdot f_{n+1}(y) \right) & \text{if } y \in S(W \oplus U^n) \text{ and } n \text{ is odd;} \\ \left(t \cdot f_{n+1}(y), \sqrt{1-t^2} \cdot f_n(y) \right) & \text{if } y \in S(W \oplus U^n) \text{ and } n \text{ is even.} \end{cases}$$

We leave it to the reader to check that f is well defined, continuous and equivariant. Thus we also have a G-map $S(W \oplus U^\infty) \simeq T \to S(W \oplus W)$. Now consider the G-vector space

$$l_2(U) := \{(u_n \colon n \in \mathbb{N}) \colon u_n \in U, \; \sum_{n=1}^\infty \|u_n\|^2 < \infty\}.$$

This is a G-Hilbert space with the usual l_2-scalar product. The continuous inclusion $S(W \oplus U^\infty) \hookrightarrow S\big(W \oplus l_2(U)\big)$ induces isomorphisms between the homotopy groups of $S(W \oplus U^\infty)^H$ and $S\big(W \oplus l_2(U)\big)^H$ for all subgroups H of G (distinguish between $U^H = 0$ and $U^H \neq 0$). Now Theorem 2.5 shows that both spheres are G-homotopy equivalent. So there exists a G-map

$$S\big(W \oplus l_2(U)\big) \simeq S(W \oplus U^\infty) \to S(W \oplus W)$$

as required. This proves Theorem 3.1b). $\square$

Proof of Corollary 3.2:

The space $W = V^K$ is an infinite-dimensional normed (H/K)-vector space with $W^{H/K} = V^H = 0$. Therefore SW is a contractible fixed point free (H/K)-ANR and Theorem 3.1a) implies $\mathcal{B}_{SW}$-genus$(SW) = \infty$ where $\mathcal{B}_{SW}$ consists of all (H/K)-orbits on SW. We want to show $\mathcal{A}_{SV}$-genus$(SV) = \infty$. Suppose not, so that there exist $G/H_1, \ldots, G/H_k \in \mathcal{A}_{SV}$ and a G-map

$$SV \to G/H_1 * \ldots * G/H_k =: X.$$

This induces an (H/K)-map $SW \to X^K$. Now all isotropy groups (with respect to H/K) appearing in X^K are also isotropy groups in SW, that is $\mathcal{B}_{X^K} \subset \mathcal{B}_{SW}$. This and the monotonicity of the genus imply the contradiction

$$\infty = \mathcal{B}_{SW}\text{-genus}(SW) \leq \mathcal{B}_{SW}\text{-genus}(X^K) \leq \mathcal{B}_{X^K}\text{-genus}(X^K) < \infty.$$

The last inequality is due to the compactness of X^K; see Proposition 2.16a). $\square$

Proof of Corollary 3.3:

We only have to prove the equivalence of (i) and (ii); the rest is a consequence of Theorem 3.1a) and Propositions 2.16 and 2.17. Let p be a prime and G_p a p-Sylow subgroup of G, that is $G_p = \pi^{-1}(W_p)$ with W_p a p-Sylow subgroup of the Weyl group $W = NT/T$ and $\pi\colon NT \to W$ the projection. Given a closed subgroup H of G it suffices to prove that p divides $\chi(G/H)$ iff G_p is not subconjugate to H. Suppose first G_p is subconjugate to H. We may assume $G_p \subset H$. The multiplicativity of the Euler characteristic (see [Sp], Theorem 9.3.1) applied to the fibration $H/G_p \subset G/G_p \to G/H$ yields $\chi(G/H) \cdot \chi(H/G_p) = \chi(G/G_p)$. Similarly we obtain $\chi(G/G_p) \cdot \chi(G_p/T) = \chi(G/T)$. Now $\chi(G/T) = \chi\big((G/T)^T\big) = |W|$ and similarly $\chi(G_p/T) = |W_p|$, hence, $\chi(G/G_p)$ is not divisible by p. Next suppose G_p is not subconjugate to H. If even the maximal torus T is not subconjugate to H then $\chi(G/H) = \chi\big((G/H)^T\big) = 0$. On the other hand, if T is subconjugate to H we may assume $T \subset H$. In this case the p-Sylow subgroup $W_p = \pi(G_p)$ is not subconjugate to $\pi(H \cap NT) = (H \cap NT)/T$. Therefore

$$\chi(G/H) = \chi(G/T)/\chi(H/T) = |W|/|(H \cap NT)/T|$$

is divisible by p.   □

Proof of Theorem 3.6:

According to Corollaries 3.13 and 3.18 there exists a fixed point free representation W_0 of G such that for every finite-dimensional fixed point free G-CW-complex X there exists a G-map $X \to SW_0$. Set $W = W_0 \oplus W_0$. Given an infinite-dimensional G-CW-complex X let X_n denote the n-skeleton. If $X^G = \emptyset$ we have G-maps $X_n \to SW_0$. As in the proof of Theorem 3.1b) consider the infinite telescope

$$T = \left(\coprod_{n \geq 1} X_n \times [0,1] \right) \Big/ \sim$$

which is G-homotopy equivalent to X. Now one defines a G-map $X \simeq T \to SW$ as in the proof of 3.1b). If X is a fixed point free G-ANR then it is G-homotopy equivalent to a G-CW-complex Y with $Y^G = \emptyset$ (see Theorem 2.4). Thus we also have a G-map $X \simeq Y \to SW$. So $W = W_0 \oplus W_0$ is a representation with the properties stated in Theorem 3.6.   □

Proof of Theorem 3.4a):

Take W as in Theorem 3.6. If U is any normed G-vector space with $U^G = 0$ then there exists a G-map $SU \to SW$. Then for $V = W \oplus U$

$$\mathcal{A}_{SV}\text{-cat}(SV) \leq 2 \cdot \mathcal{A}_{SW}\text{-cat}(SW) < \infty$$

because of the Propositions 2.18 and 2.13a).   □

Proof of Theorem 3.4b):

If X is a fixed point free G-ANR then Theorem 3.6 yields a G-map $X \to SW$. If moreover $\mathcal{A}_{SW} \subset \mathcal{A}_X$ then by the monotonicity of the genus

$$\mathcal{A}_X\text{-genus}(X) \leq \mathcal{A}_{SW}\text{-genus}(X) \leq \mathcal{A}_{SW}\text{-genus}(SW) < \infty.$$

□

Proof of Theorem 3.5, (i)⇒(ii):

Suppose (ii) is false and let $f: SV \to SW$ be a G-map between representation spheres with $W \subsetneq V$ and $SW^G = \emptyset$. Setting $U = W^\perp \subset V$ one constructs a G-map $S(W \oplus l_2(U)) \to S(W \oplus W)$ as in the proof of Theorem 3.1b). This contradicts (i).
□

Proof of Theorem 3.5, (ii)⇒(iii):

This is precisely Proposition 3.8.   □

Proof of Theorem 3.5, (iii)⇒(i):

Let $SV \to SW$ be a G-map as in 3.5(i). Then $U = V \oplus W$ is a fixed point free G-Hilbert space and (using 2.18)

$$\mathcal{A}_{SU}\text{-cat}(SU) \leq 2 \cdot \mathcal{A}_{SU}\text{-cat}(SW) = 2 \cdot \mathcal{A}_{SW}\text{-cat}(SW).$$

Theorem 3.1a) implies $\mathcal{A}_{SU}\text{-cat}(SU) = \infty$, hence $\dim W = \infty$ by 2.13a).   □

Proof of Theorem 3.7a), (i)⇒(ii):

Suppose (ii) is false, i.e. either the connected component G_0 of G is not a torus or the factor group G/G_0 is not solvable. In the first case let TZ_0 be the maximal torus of the center Z_0 of G_0 and set $\Gamma_0 = G_0/TZ_0$. Γ_0 is a compact connected semisimple Lie group (see [Ho], Theorem XIII.1.3), hence there exists a fixed point free representation W_0 of Γ_0 and a Γ_0-map $f: S(W_0 \oplus W_0) \to SW_0$; see Corollary 3.18. We consider W_0 as representation of G_0 via the projection $G_0 \to \Gamma_0$. Setting $V := \text{ind}_{G_0}^G W$ we obtain a G-map $\text{ind}_{G_0}^G f: S(V \oplus V) \to SV$. V is a fixed point free representation of G which is not admissible.

Now assume that $\Gamma = G/G_0$ is not solvable. It suffices to construct a fixed point free representation of Γ which is not admissible. Since Γ is not solvable it contains a subgroup Δ which has no subgroup of prime index. Theorem 3.6 (or Corollary 3.13) yields a Δ-map $f: S(W \oplus W) \to SW$ as required. Now we pass to the induced map $\text{ind}_\Delta^\Gamma f$ which shows that $\text{ind}_\Delta^\Gamma W$ is not admissible.   □

Proof of Theorem 3.7a), (ii)⇒(i):

Let G_0 be a torus and $\Gamma = G/G_0$ be solvable. Consider a fixed point free representation V of G. We have to show that every G-map $f: \overline{\mathcal{O}} \to SV^{k-1}$, where $\mathcal{O}$ is an open bounded invariant neighborhood of 0 in V^k, has a zero in $\partial\mathcal{O}$. We consider first the special case G finite. Let $p_1^{m_1} \cdot \ldots \cdot p_r^{m_r}$ be the order of G where $p_1, \ldots, p_r$ are prime numbers and $m_i \geq 1$. We use induction on $m = \sum_{i=1}^r m_i$. If

$m = 1$ then $G \cong \mathbb{Z}/p$ and every fixed point free representation is admissible because a G-map $f: \overline{\mathcal{O}} \to V^{k-1}$ with $f(\partial\mathcal{O}) \subset V^{k-1} - 0$ leads to the contradiction

$$0 = \deg(i \circ f, \mathcal{O}, 0) \equiv 1 \pmod{p};$$

here $i: V^{k-1} \hookrightarrow V^k$ is the inclusion. The above congruence is well known. According to [Ste] the first reference is probably a paper by Eilenberg [Ei]. An elementary proof can be found in [Ba4]. Now let $m \geq 2$ and consider a fixed point free representation V of G and a G-map $f: \overline{\mathcal{O}} \to V^{k-1}$ as above. Since G is solvable there exists a normal subgroup H of G different from $\{e\}$ or G. Now either $V^H = 0$ or $V^H \neq 0$. In the first case V is a fixed point free representation of H. By the induction hypothesis f has a zero in $\partial\mathcal{O}$. In the second case set $W := V^H$ which is a representation of G/H since H is a normal subgroup of G. Observe that $f(\operatorname{clos}(\mathcal{O} \cap W^k)) \subset W^{k-1}$ because equivariant maps preserve fixed point sets. The restriction $\operatorname{clos}(\mathcal{O} \cap W^k) \xrightarrow{f} W^{k-1}$ is a G/H-map. Moreover $W^{G/H} = V^G = 0$, so W is fixed point free and the induction hypothesis yields a zero of f in $\partial(\mathcal{O} \cap W^k)$.

Thus we have proved that for a finite solvable group every fixed point free representation is admissible. We consider now the general case where G_0 is a torus and $\Gamma = G/G_0$ is solvable. Let $f: \overline{\mathcal{O}} \to V^{k-1}$ be a G-map as above. According to [Di], Proposition IV(3.6), G is the limit of finite subgroups. Therefore there exists a finite subgroup H of G with $V^H = V^G = 0$. H is solvable as an extension of an abelian group by a solvable group. Now V is a fixed point free representation of H, hence admissible. Thus f must have a zero in $\partial\mathcal{O}$. Another simple argument which avoids the reference to [Di] is to distinguish between $V^{G_0} = 0$ or $V^{G_0} \neq 0$. We leave this to the reader.   □

Proof of Theorem 3.7b):

Let V be a representation of G such that there exist subgroups $K \subset H$ of G with K normal in H and H/K solvable satisfying $V^H = 0$ and $V^K \neq 0$. Then a G-map $f: \overline{\mathcal{O}} \to V^{k-1}$, $\mathcal{O} \subset V^k$ as usual, induces an H/K-map $f^K: \operatorname{clos}(\mathcal{O} \cap W^k) \to W^{k-1}$ where $W = V^K$. Since $W^{H/K} = V^H = 0$ part a) shows that W is an admissible representation of H/K, hence f has a zero in $\partial(\mathcal{O} \cap W^k)$.

Next let V be a representation of G such that for every pair $K \subset H$ of subgroups of G with K normal in H and K/H solvable, $V^H = 0$ implies $V^K = 0$. We claim that V is not admissible. We first define a possible degree function $d: \psi(G) \to \mathbb{Z}$ of a G-map $SV^2 \to SV^2$ as follows:

$$d(H) := \begin{cases} 0 & \text{if } V^H \neq 0, \\ 1 & \text{if } V^H = 0. \end{cases}$$

We want to apply Proposition 3.10 to show that d is the degree function of some map $f: SV^2 \to SV^2$, that is $\deg(f^H) = d(H)$ for every $(H) \in \psi(G)$. Observe that V^2 can be considered as a complex representation of G. We have to check the assumptions of 3.10. First of all, d is continuous: If $(H_i) \to (H)$ then $V^{H_i} = V^H$ for almost all i because $\dim V < \infty$. Hence $d(H_i) = d(H)$ for almost all i. Next, if K is a normal subgroup of H and H/K is a torus then $d(H) = d(K)$ by our assumption on V and by definition of d. Now we come to the congruences 3.10(ii) for subgroups H of G with $WH = NH/H$ finite:

$$\sum_{(L)} m(H,L) \cdot d(L) \equiv 0 \quad (\bmod\ |WH|).$$

Here (L) runs through all NH-conjugacy classes of subgroups $H \subset L \subset NH$ such that L/H is finite cyclic. This implies $d(L) = d(H)$ for all L in the sum. Thus the congruence is trivially true if $d(H) = 0$. And if $d(H) = 1$ then the congruence holds because it is the congruence for the degree function of the identity map $SV^2 \to SV^2$. Finally, conditions 3.10(iii) and (iv) are true by definition of d.

Thus there exists a G-map $f \colon SV^2 \to SV^2$ with $\deg(f^H) = 0$ if $V^H = 0$. Now Lemma 3.9 yields a G-map $SV^3 \to SV^2$ as required. This shows that V is not admissible and finishes the proof of Theorem 3.7. □

3.5 Related results and examples

First we illustrate the results of §3.2 by some examples. All representations are fixed point free in this section.

3.19 Example:
Consider the cyclic group $G = \mathbb{Z}/pq$ with two different primes p and q. Any (fixed point free) real representation V is admissible according to Theorem 3.7, i.e. there cannot exist a G-map $SV^k \to SV^{k-1}$. If $\dim V^{\mathbb{Z}/p} \geq 2$ and $\dim V^{\mathbb{Z}/q} \geq 2$ then there exists a nullhomotopic G-map $SW \to SW$ with $W := V^{\mathbb{Z}/p} \oplus V^{\mathbb{Z}/q} \subset V$; see the proof of Proposition 3.8 for $\mathbb{Z}/pq$. Now Lemma 3.9 yields a G-map $SV \to SW$. And if $\dim V^{\mathbb{Z}/p} \leq 1$ or $\dim V^{\mathbb{Z}/q} \leq 1$ then there cannot exist a G-map $SV \to SW \subsetneq SV$. This is clear if both p and q are odd primes because then all fixed point free representations are even-dimensional. Hence, $\dim V^{\mathbb{Z}/p} \leq 1$ implies $V^{\mathbb{Z}/p} = 0$ (because $V^{\mathbb{Z}/p}$ is a fixed point free representation of $\mathbb{Z}/q$), so that $\mathbb{Z}/p$ acts freely on SV. The non-existence of a G-map $SV \to SW$ in the case $q = 2$ and $\dim V^{\mathbb{Z}/p} = 1$ requires a simple extra treatment using the congruences of Proposition 3.10 for G-maps $SW \to SW$. (These congruences also hold for real representations.) Suppose now that V is a (possibly infinite-dimensional) normed G-vector space with $V^G = 0$. Then Corollary 3.2 and Proposition 2.17 imply the following:

$$\mathcal{A}_{SV}\text{-cat}(SV) = \infty \quad \text{if } \dim V^{\mathbb{Z}/p} \in \{0,\infty\} \text{ or } \dim V^{\mathbb{Z}/q} \in \{0,\infty\};$$

and

$$\mathcal{A}_{SV}\text{-cat}(SV) < \infty \quad \text{if } 2 \leq \dim V^{\mathbb{Z}/p} < \infty \text{ and } 2 \leq \dim V^{\mathbb{Z}/q} < \infty.$$

These statements cover all possible V if p and q are odd primes. If $q = 2$, $\dim V^{\mathbb{Z}/p} = 1$, $1 \leq \dim V^{\mathbb{Z}/q} < \infty$ and $\dim V = \infty$ we do not know whether $\mathcal{A}_{SV}\text{-cat}(SV)$ is infinite or not. For the genus the situation is clear:

$$\mathcal{A}_{SV}\text{-genus}(SV) < \infty \quad \text{iff } 1 \leq \dim V^{\mathbb{Z}/p} < \infty \text{ and } 1 \leq \dim V^{\mathbb{Z}/q} < \infty.$$

Another amusing consequence of the existence of G-maps as above together with the weak monotonicity of the category (cf. Proposition 2.18) is the following concrete upper bound for the category. Let $W \cong \mathbb{C}^2$ be the representation of G as in the beginning of the proof of Proposition 3.8; and let U be any finite-dimensional G-vector space with a free action of G. Then for any $n \geq 1$ and $\mathcal{A} = \mathcal{A}_{S(U \oplus W)}$

$$\operatorname{cat}\left(S(U \oplus W^n)/G\right) \leq \mathcal{A}\text{-cat}\left(S(U \oplus W^n)\right) \leq \begin{cases} 4n & \text{if } n \text{ is even;} \\ 4(n+1) & \text{if } n \text{ is odd;} \end{cases}$$

independently of the dimension of U. For even n one can show that equality holds. Recall that $\operatorname{cat}(SU/G) = \dim U$.

$G = \mathbb{Z}/pq$ is the easiest case where one knows precisely whether or not a G-map $SV \to SW \subsetneq SV$ exists. Slightly more complicated is the case where G is an arbitrary finite abelian group. But instead of generalizing Example 3.19 we look at a completely different group important in applications.

3.20 Example:
   Let $G = SO(3)$ be the group of rotations in $\mathbb{R}^3$. The irreducible representations of G are the spaces V_l, $l = 0, 1, 2, \ldots$, of *spherical harmonics* of order l. Elements of V_l are the homogeneous polynomials $f = f(x_1, x_2, x_3)$ in three variables of degree l satisfying $\Delta f = 0$; Δ is the Laplace operator on $\mathbb{R}^3$. The dimension of V_l is $2l + 1$. We refer the reader to the book [BröD], §II.5.3, for a discussion of the V_l. $SO(3)$ has (up to conjugacy) three maximal subgroups: the orthogonal group $O(2)$, the icosahedral group I and the octahedral group O. I is isomorphic to the alternating group A_5, in particular I is simple. O is isomorphic to the symmetric group S_4, so O is solvable. The other subgroups of $SO(3)$ are the tetrahedral group T which is isomorphic to A_4 and contained in O and I and the subgroups of $O(2)$, i.e. cyclic and dihedral subgroups. Therefore if W_0 is a complex representation of $SO(3)$ with $\dim W_0^H \geq 1$ for $H = O(2), I, O$ then any finite-dimensional fixed point free G-ANR X can be mapped equivariantly into SW_0 according to Corollary 3.18. For the irreducible representations V_l the dimensions of the fixed point spaces are as follows (see [IG], Theorem 3.2):

$$\dim V_l^{O(2)} = \begin{cases} 0 & \text{if } l \text{ is odd,} \\ 1 & \text{if } l \text{ is even;} \end{cases}$$
$$\dim V_l^O = [l/4] + [l/3] + [l/2] - l + 1;$$
$$\dim V_l^I = [l/5] + [l/3] + [l/2] - l + 1.$$

Here $[x]$ is the greatest integer less than or equal to x. Therefore

$$\dim V_l^O \geq 1 \quad \text{iff } l = 0, 4, 6, 8, 9, 10 \text{ or } l \geq 12;$$
$$\dim V_l^O \geq 2 \quad \text{iff } \dim V_{l-12}^O \geq 1;$$
$$\dim V_l^I \geq 1 \quad \text{iff } l = 0, 6, 10, 12, 15, 18, 20, 24 \text{ or } l \geq 30;$$
$$\dim V_l^I \geq 2 \quad \text{iff } \dim V_{l-30}^I \geq 1.$$

Now it is trivial to determine the possible W_0 as above. The lowest dimensional one is $W_0 = V_6 \oplus V_6$, hence $\dim W_0 = 26$. Any finite-dimensional fixed point free G-ANR X can be mapped into SW_0. And an infinite-dimensional G-ANR X with $X^G = \emptyset$ can be mapped into $S(W_0 \oplus W_0)$. Thus $(V_6)^4$ is a representation of $SO(3)$ whose existence has been claimed in Theorems 3.4 and 3.6.

It is more complicated to determine all representations V such that there exists a G-map $SV \rightarrow SW \subsetneq SV$. This can certainly not happen if V contains only summands V_l with odd l because $\dim V_l^{SO(2)} = 1$ for any l, $\dim V_l^{O(2)} = 0$ for odd l and $SO(2)$ is a normal subgroup of $O(2)$. The existence of a G-map $SV \rightarrow SW \subset SV$ yields a $\mathbb{Z}/2$-map $SV^{SO(2)} \rightarrow SW^{SO(2)}$ which implies $\dim W^{SO(2)} = \dim V^{SO(2)}$, hence $W = V$. This argument does not work if V contains some V_l with l even. On the other hand, not so many cases remain where $\dim V_l^O \leq 1$ or $\dim V_l^I \leq 1$. Many of these cases can be treated using Lemma 3.9 and Proposition 3.10. We leave this to the interested reader as well as the classification of those infinite-dimensional normed G-vector spaces V with $\mathcal{A}_{SV}$-cat$(SV) = \infty$ or $\mathcal{A}_{SV}$-genus$(SV) = \infty$. Again, a few cases will remain, at least if one only applies the results established in this chapter.

Finally, we classify the admissible representations V completely using Theorem 3.7b). We have to find subgroups $K \subset H$ of $SO(3)$ with K normal in H, H/K solvable and $V^H = 0 \neq V^K$. It suffices to consider those pairs $K \subset H$ where H/K is finite. The only such pairs are $SO(2) \subset O(2)$, $T \subset O$, $\mathbb{Z}/n \subset H \subset O(2)$, $D_n \subset D_{2n}$ for $n > 2$ and $D_2 \subset O$. Here D_n is the dihedral group of order $2n$. The only cases where $V_l^H = 0 \neq V_l^K$ are l odd with $H = O(2)$, $K = SO(2)$, and $l = 2$ with $H = O$ and $K = D_2$. Therefore V is admissible iff it contains only summands V_l with odd l or it contains only V_1, V_2, V_3, V_5, V_7 or V_{11}. In other words, if V contains at least one V_l with even $l \neq 2$ or if it contains V_2 and at least one V_l with $l = 9$ or $l \geq 12$ then V is not admissible.

So far our results distinguish between p-groups, solvable groups and non-solvable groups, and for applications the case of p-groups is the nicest. But there are also interesting problems with the p-group case. We consider first an example.

3.21 Example:

Let $G = \mathbb{Z}/4 \subset S^1$ act on $V \cong \mathbb{C}^2$ via scalar multiplication and on $W \cong \mathbb{R}^3$ via the antipodal map. So V is a fixed point free representation and the isotropy group on SW is $\mathbb{Z}/2$. The map

$$SV \ni (z_1, z_2) \mapsto \left((z_1 + \bar{z}_2) \cdot (z_1 - \bar{z}_2), |z_1 + \bar{z}_2| - |z_1 - \bar{z}_2| \right) \in W - 0$$

followed by the retraction $W - 0 \rightarrow SW$ is a $\mathbb{Z}/4$-map $SV \rightarrow SW$. Its homotopy class generates $\pi_3(S^2) \cong \mathbb{Z}$. Setting $\mathcal{A} = \{G/H : H \subsetneq G\}$ the existence of this G-map shows

$$\mathcal{A}\text{-genus}(SV) \leq \mathcal{A}\text{-genus}(SW) = 3 < \dim V$$

and by the weak monotonicity property of $\mathcal{A}$-cat (cf. Proposition 2.18)

$$\mathcal{A}\text{-cat}(S(V \oplus W)) \leq 2 \cdot \mathcal{A}\text{-cat}(SW) = 6 < \dim(V \oplus W).$$

This example is typical for p-groups which have an element of order p^2.

3.22 Theorem:
  Let G be a p-group containing an element of order p^2. Then there exists a G-map $SV \to SW$ between fixed point free representations V and W such that $\dim W < \dim V$.

Proof:
  It suffices to consider the case $G = \mathbb{Z}/p^2$. Let $\zeta \in G \subset S^1$ act on $V_j \cong \mathbb{C}$ via scalar multiplication with ζ^j. According to [Mar], Theorem 2.8, there exists a G-map $f_1 \colon SV_1^2 \to SV_p^2$ of degree 0. If $h_t \colon SV_1^2 \to SV_p^2$ is a homotopy between $h_0 = \text{const}$ and $h_1 = f_1$ we define

$$f_2 \colon G * SV_1^2 \to SV_p^2, \qquad f_2([1-t, g, t, v]) := g \cdot h_t(g^{-1}v).$$

This is well defined, continuous and equivariant since h_1 is equivariant and h_0 is constant. The join $f_2 * f_2$ induces a G-map

$$f_3 \colon G * G * SV_1^4 \cong G * SV_1^2 * G * SV_1^2 \xrightarrow{f_2 * f_2} SV_p^4.$$

Finally there exists a G-map $SV_1 \to G * G$ since SV_1 is a free G-space and

$$\{G\}\text{-genus}(SV_1) \leq \{G\}\text{-cat}(SV_1) = \text{cat}(SV_1/G) = 2.$$

Of course one can also define such a map explicitely. Therefore we obtain a G-map

$$SV_1^5 \to G * G * SV_1^4 \xrightarrow{f_3} SV_p^4$$

as required.   □

  So even for p-groups it is very well possible that there exist G-maps $SV \to SW$ between fixed point free representation spheres with $\dim W < \dim V$. Of course, W cannot be a subrepresentation of V. As in Example 3.21 one sees that $\mathcal{A}$-genus$(SV) < \dim V$ and $\mathcal{A}$-cat$\big(S(V \oplus W)\big) < \dim(V \oplus W)$. For the computation of the category and the genus it is important to know how small $\dim W$ can be if $\dim V$ is given. So we are interested in the following *Borsuk-Ulam function* $b_G \colon \mathbb{N} \to \mathbb{N}$.

  For $n \in \mathbb{N}$ we define $b_G(n)$ to be the maximal natural number k such that the existence of a G-map $SV \to SW$ between fixed point free representation spheres with $\dim V \geq n$ implies $\dim W \geq k$.

In other words, if there is a G-map $SV \to SW$ as above then $\dim W \geq b_G(\dim V)$, and the Borsuk-Ulam function b_G is maximal with respect to this property. The computation of b_G is difficult except for $G = (\mathbb{Z}/p)^k$ or $G = (S^1)^k$. For these groups we have $b_G(2n) = 2n$ and

$$b_G(2n+1) = \begin{cases} 2n+1 & \text{if } G = (\mathbb{Z}/2)^k, \\ 2n+2 & \text{if } G \text{ is a } (p\text{-})\text{torus } (p \text{ odd}). \end{cases}$$

The computation of b_G is difficult. So far the following is known.

3.23 Theorem:

a) *For $G = \mathbb{Z}/4$ we have: $b_G(1) = 1$, $b_G(2) = 2$ and if $n \geq 2$*

$$b_G(2n) = \begin{cases} n+1 & \text{for } n \equiv 0, 2 \pmod 8; \\ n+2 & \text{for } n \equiv 1, 3, 4, 5, 7 \pmod 8; \\ n+3 & \text{for } n \equiv 6 \pmod 8. \end{cases}$$

$$b_G(2n-1) = \begin{cases} n+1 & \text{for } n \equiv 0, 1, 2, 3 \pmod 8; \\ n+2 & \text{for } n \equiv 6, 7 \pmod 8. \end{cases}$$

If $n \equiv 4, 5 \pmod 8$ then $n + 1 \leq b_G(2n-1) \leq n + 2$.

b) *For $G = \mathbb{Z}/p^2$ with p an odd prime we have: $b_G(1) = 1$, $b_G(2) = 2$ and if $n \geq 2$*

$$b_G(2n) \in \left\{ 2 \left\langle \frac{n-1}{p} \right\rangle + 2 \, , \, 2 \left\langle \frac{n-1}{p} \right\rangle + 4 \right\}$$

and $b_G(2n-1) = b_G(2n)$ because there are no odd-dimensional representations without fixed points. Here $\langle x \rangle$ is the least integer greater than or equal to x. □

A proof of part a) can be found in [Ba3]. It is a consequence of a result of Stolz [Sto]. In [Ba2] the lower bound in part b) is proved; see also Corollary 5.9 and Remark 5.10. The upper bound in b) is proved in [Me]. Section 5.3 contains computations for the cyclic groups $G = \mathbb{Z}/p^k$ including estimates for the category and genus of representation spheres. Earlier results in this direction are due to Munkholm [Mun] and Vick [Vick].

Closely related to the question whether spheres of infinite-dimensional normed G-vector spaces have infinite category is the question whether b_G is bounded. In fact, for finite G these properties are equivalent: b_G is unbounded if and only if G is a p-group; see [Ba3]. For a general compact Lie group Theorem 3.5 tells us that b_G is bounded if G is not p-toral. We expect that the converse is also true. Observe that b_G is weakly monotone ($b_G(n) \leq b_G(m)$ if $n \leq m$), subadditive ($b_G(m+n) \leq b_G(m) + b_G(n)$) and satisfies $b_G(n) \leq n$ if there exists an n-dimensional fixed point free representation. This implies that $a_G := \lim_{n \to \infty} b_G(n)/n$ exists and $0 \leq a_G \leq 1$. Furthermore, $b_G(n) \geq a_G \cdot n$ for all n. So far the following is known about a_G:

3.24 Theorem:

a) $a_G = 1$ if $G = (\mathbb{Z}/p)^k$ or $G = (S^1)^k$.

b) $a_G = 1/p^k$ if $G = \mathbb{Z}/p^{k+1}$.

c) $a_G \leq 1/p^k$ if G is p-toral and the (finite) factor group G/G_0 has an element of order p^{k+1}.

d) $a_G = 0$ if G is not p-toral.

Proof:

a) This is a consequence of known versions of the Borsuk-Ulam theorem; cf. [Ba4] or Remark 5.6.

b) The results of [Ba2] or §5.3 show that $a_G \geq 1/p^k$ for $G = \mathbb{Z}/p^{k+1}$. In order to see the reverse inequality we argue by induction on k as follows. The inequality $a_G \leq 1/p^k$ holds for $k = 0, 1$ by Theorem 3.23. Now consider $H := \mathbb{Z}/p^k \subset G = \mathbb{Z}/p^{k+1}$ and assume $a_H \leq 1/p^{k-1}$. We write $V_i \cong \mathbb{C}$ for the representation of G given by the character $\zeta \mapsto \zeta^i$, $i = 0, 1, \ldots, p^{k+1} - 1$. Similarly we write $W_i \cong \mathbb{C}$ for the corresponding representation of H, $i = 0, 1, \ldots, p^k - 1$. Set $g := p^k = |G|/p$ and $h := p^{k-1} = |H|/p$. Thus by induction, $a_H = 1/h$ and $b_H(hn) = n + o(n)$ as $n \to \infty$. This just means $(b_H(n) - n)/n \to 0$ as $n \to \infty$. We use similar notation below. Thus there exist H-maps

$$f_n \colon SW_1^{hn} \xrightarrow{\ H\ } SW_h^{n+o(n)}$$

if n is sufficiently large. Now one checks that $\operatorname{ind}_H^G(W_i) \cong V_i^p$, hence passing to the induced map (see the proof of Proposition 3.8) we obtain a G-map

$$\operatorname{ind}_H^G(f_n) \colon SV_1^{gn} \xrightarrow{\ G\ } SV_h^{pn+o(n)}.$$

Next one checks that $K := \mathbb{Z}/h \subset G$ acts trivially on SV_h, and $G/K \cong \mathbb{Z}/p^2$ acts freely on $SV_h = SV_h^K$. Since we know that $a_{G/K} \leq 1/p$ (this is the case $k = 1$) there exist G-maps

$$SV_h^{pn+o(n)} \xrightarrow{\ G\ } SV_g^{n+o(n)}$$

for sufficiently large n. Composed with the maps $\operatorname{ind}_H^G(f_n)$ we obtain G-maps

$$SV_1^{gn} \xrightarrow{\ G\ } SV_g^{n+o(n)}$$

for sufficiently large n. This implies $b_G(gn) \leq n + o(n)$ if n is big enough. Hence,

$$a_G = \lim_{n \to \infty} b_G(gn)/gn \leq 1/g = 1/p^k$$

as claimed.

c) Observe that $b_G \leq b_H$, hence $a_G \leq a_H$, if there exists an epimorphism $G \to H$. This implies $a_G \leq a_\Gamma$ where $\Gamma = G/G_0$. Moreover, $a_G \leq a_H$ if H is a nontrivial subgroup of G of finite index, as one can see using induced representations. In particular, $a_\Gamma \leq 1/p^k$ if γ contains an element (a cyclic subgroup) of order p^{k+1} by part b).

d) This is clear since b_G is even bounded if G is not p-toral according to Theorem 3.1b).   □

One is tempted to conjecture that $a_G = p/\exp(G/G_0)$ if G is a p-toral group (but not a torus). Recall that the exponent $\exp(\Gamma)$ of the finite group $\Gamma = G/G_0$ is the least common multiple of the orders of the elements of Γ. So 3.24c) says $a_G \leq p/\exp(G/G_0)$. If this conjecture was true we would have a minimax-description of a_G:

$$a_G \overset{?}{=} \min_{\gamma \in \Gamma - e} \max\{1/p^l : \gamma^{p^{l+1}} = e\} .$$

There are various generalizations of the Borsuk-Ulam theorem which are related to the results of this chapter. We refer the reader to the survey [Ste] and its extensive bibliography. Some recent papers which are relevant in this context and not listed there are [ClP2], [IM], [Mar], [To], [Wang] and [Was]. These papers contain results about the non-existence of G-maps between certain spaces and/or computations of the possible degrees of G-maps between representation spheres of the same dimension. The paper [Was] is different in spirit insofar only isovariant G-maps are being considered (a G-map $f: X \to Y$ is isovariant if $G_{f(x)} = G_x$ for every $x \in X$). Also of interest are the papers [AtT] and [LW] about the existence of G-maps $SV \to SW$ with degree prime to the order of G.

Chapter 4

The length of G-spaces

4.1 Introduction

The computation of the $\mathcal{A}$-category and the $\mathcal{A}$-genus falls naturally into two parts. Upper bounds are obtained by (more or less) explicit constructions of certain G-coverings or G-maps. Chapter 3 was devoted to proving results in this direction. To find lower bounds for the category and the genus one has to prove the non-existence of certain coverings or maps. At this point algebraic topology enters.

Naturally we try to find an equivariant version of the cup-length. To do this we need to know a bit about equivariant cohomology theories with an interior product (the cup-product). We collect basic notation and examples in §4.2. In §4.3 we first indicate the proper version of the equivariant cup-length. As far as lower bounds for category and genus are concerned we could stop any further discussion and proceed directly to computations. But for later applications in critical point theory we need a version which satisfies at least the properties of $\mathcal{A}$-cat$_M$ and $\mathcal{A}$-genus. More precisely, it should have all the properties of the cohomological index theories introduced by Fadell, Husseini and Rabinowitz in [FaR1,2], [FaHR], [FaH1,2]. Motivated by the definition of the category and of the equivariant cup-length we define the "length". This is a cohomological index theory which coincides with the one of Fadell et. al. in the cases $G = \mathbb{Z}/2$ and $G = S^1$. For other groups it is different and provides more and better information, already if G is a (p-)torus. The length as defined in §4.3 is a modified version of the one defined and applied in several papers of Clapp, Puppe and the author ([ClP2], [BCP], [BC1]). As the category and genus it depends on a set $\mathcal{A}$ of G-spaces. It also depends on the choice of a multiplicative equivariant cohomology theory h^* and on the choice of an ideal I of the coefficient ring $h^*(\text{pt})$. The length will be denoted by $(\mathcal{A}, h^*, I)$-length and — since this notation is somewhat awkward — simply by ℓ if $(\mathcal{A}, h^*, I)$ are understood.

In §4.4 we prove a number of elementary properties of the length which partly resemble those of the category and the genus. We do not try to give an axiomatic description of the length similar to those for $\mathcal{A}$-cat$_M$ and $\mathcal{A}$-genus in Propositions 2.12 and 2.15. However we do try to state simple properties first and deduce others later. The results of §4.4 do not depend on a particular choice of h^* because they only use the Eilenberg-Steenrod axioms (not the dimension axiom) plus the general properties of the interior product.

One can certainly prove more results if one specializes the group G and the cohomology theory h^*. Dictated by later needs we collect some results of this type in §4.5. There we work with Borel cohomology H_G^*. This is in a certain sense the most basic equivariant cohomology theory directly related to singular (or Alexander-Spanier) cohomology. We shall refer to H_G^* whenever we need an example. Due to its computability H_G^* also plays an important role in applications.

4.2 Equivariant cohomology theories

A *G-equivariant cohomology theory* h^* is, as usual, a sequence of abelian group-valued contravariant functors h^n, $n \in \mathbb{Z}$, defined on the category of pairs (X, X') of *G*-spaces $X' \subset X$ and continuous equivariant maps. h^* satisfies the following equivariant versions of the Eilenberg-Steenrod axioms:

(i) *Homotopy invariance:* If $f, g \colon (X, X') \to (Y, Y')$ are *G*-homotopic then

$$f^* = g^* \colon h^*(Y, Y') \longrightarrow h^*(X, X').$$

(ii) *Exactness:* For every pair (X, X') of *G*-spaces there is a natural exact sequence

$$\ldots \xrightarrow{\delta} h^n(X, X') \xrightarrow{j^*} h^n(X) \xrightarrow{i^*} h^n(X') \xrightarrow{\delta} h^{n+1}(X, X') \xrightarrow{j^*} \ldots$$

where i and j denote inclusions.

(iii) *Excision:* If X' and X'' are invariant subspaces of X whose interiors cover X then the inclusion $(X'', X' \cap X'') \hookrightarrow (X, X')$ induces an isomorphism

$$h^*(X, X') \xrightarrow{\cong} h^*(X'', X' \cap X'').$$

As a consequence of these axioms one obtains the exact sequence of a triple and the Mayer-Vietoris sequence of an excisive couple (cf. [EiS] or [Sw]). In order to obtain a lower bound for the category and the genus we shall also need a multiplicative structure. A *G*-equivariant cohomology theory h^* is said to be *multiplicative* if it has an interior product (the cup-product):

$$\smile \colon h^m(X, X') \times h^n(X, X'') \longrightarrow h^{m+n}(X, X' \cup X''),$$

defined (at least) for all *G*-pairs (X, X'), (X, X'') such that the interiors of X' and X'' (with respect to $X' \cup X''$) cover $X' \cup X''$. This assumption is satisfied, for example, if $\overline{X'} \cap \overline{X''} = \emptyset$ or if both X' and X'' are open *G*-subsets of X. Furthermore, $\smile$ has to be natural, bilinear, associative, commutative (up to the usual sign $(-1)^{mn}$), with unit in $h^0(X)$ and compatible with the connecting homomorphism δ. This product turns $h^*(X)$ into a ring with unit and $h^*(X, X')$ into a module over $h^*(X)$, for any *G*-pair (X, X'). The constant (equivariant) map $X \to \mathrm{pt}$ induces a ring homomorphism so that $h^*(X, X')$ is also a module over $h^*(\mathrm{pt})$. Here pt denotes the one-point space. $h^*(\mathrm{pt})$ is the *coefficient ring* of h^*.

Given a multiplicative equivariant cohomology theory h^* one can define a lower bound for the category or genus as we shall see in the next section. But for applications to critical point theory it is necessary that h^* satisfies the following continuity property: h^* is said to be *continuous* if for any closed *G*-subset X' of a metrizable *G*-space X

$$h^*(X') \cong \varinjlim h^*(U)$$

where the direct limit is taken over all open *G*-neighborhoods U of X' in X.

If h^* is continuous then the above isomorphism also holds if X' is locally closed, that is $X' = C \cap O$ is the intersection of a closed subset C of X and an open subset O of X. Namely,

$$h^*(X') \cong \varinjlim h^*(V) \cong \varinjlim h^*(U)$$

where the direct limits are taken over all open G-neighborhoods V of X' in O resp. U of X' in X. The first isomorphism holds since O is metric and X' is a closed subset of O. The second holds since the neighborhoods of X' in O are a cofinal subset of the neighborhoods of X' in X. A second consequence is that

$$h^*(X', X'') \cong \varinjlim h^*(U, V)$$

where $X'' \subset X' \subset X$ are locally closed G-subsets of X. To see this simply apply the 5-lemma.

If h^* is both multiplicative and continuous then the product can be defined if X is metrizable and X', X'' are locally closed subspaces of $X' \cup X''$. Namely, for $\xi_1 \in h^m(X, X')$ respectively $\xi_2 \in h^n(X, X'')$ there exist open G-neighborhoods U' of X' respectively U'' of X'' in $X' \cup X''$ and elements $\eta_1 \in h^m(X, U')$ respectively $\eta_2 \in h^n(X, U'')$ which restrict to ξ_i:

$$\xi_i = \eta_i | (X, X^{(i)})$$

where the restriction $|(X, X^{(i)}): h^*(X, U^{(i)}) \longrightarrow h^*(X, X^{(i)})$ is induced by the inclusion. Then we set $\xi_1 \smile \xi_2 := (\eta_1 \smile \eta_2)|(X, X' \cup X'')$. It is straightforward to check that $\xi_1 \smile \xi_2$ is well defined and that this induces a product

$$\smile : h^m(X, X') \times h^n(X, X'') \longrightarrow h^{m+n}(X, X' \cup X'')$$

with the usual properties.

We have been very careful in dealing with continuity and multiplicativity since we need to apply these properties in their full generality. An important example of an equivariant cohomology theory is Borel cohomology H_G^*. This is defined as follows (a general reference is [Di], III.1–III.2).

Borel cohomology:

Let $H^*(-; R)$ denote Alexander-Spanier cohomology with coefficients in a commutative ring R with unit ([Sp], Section 6.4). Let EG be a contractible space on which G acts freely. EG exists for any topological group and is uniquely determined up to G-homotopy. For $G = \mathbb{Z}/2 \cong \{\pm 1\}$ we may take the unit sphere of any infinite-dimensional normed vector space with the action given by scalar multiplication. For $G = S^1$ we can do the same except that the vector space should be complex. Clearly, if H is a subgroup of G then EG can also serve as a model for EH. A standard model is the infinite join $EG = G * G * \ldots$ (Milnor's construction). Another model can be obtained as follows. Let $\mathbb{R}^\infty := \bigcup_{k=1}^\infty \mathbb{R}^k$ be the union of the Euclidean spaces $\mathbb{R}^k$ with the weak topology and S^∞ the unit sphere of this pre-Hilbert space. Then we define

$$EO(n) := \{(u_1, \ldots, u_n) \in \mathbb{R}^\infty : \langle u_i, u_j \rangle = \delta_{ij} \quad \forall 1 \le i, j \le n\} \subset (S^\infty)^n$$

to be the *Stiefel manifold of orthogonal n-frames* in $\mathbb{R}^\infty$. If $n = 1$ this is just the unit sphere. The action of $A = (a_{ij}) \in O(n)$ on $u = (u_1, \ldots, u_n) \in EO(n)$ is given by

$$Au := v \qquad \text{with } v_i := \sum_{j=1}^{n} a_{ij} u_j, \ i = 1, \ldots, n.$$

Knowing $EO(n)$ for every n we also know EG because every compact Lie group G is a closed subgroup of some $O(n)$. Therefore $EO(n)$ serves as a model for EG. Since we only deal with compact Lie groups G we may assume that EG is metrizable. The orbit space EG/G is denoted by BG. The classifying spaces BG are more complicated; BG classifies principal G-bundles, i.e. bundles with fibre G over compact spaces (for example). Obviously $B\mathbb{Z}/2 = \mathbb{R}P^\infty$ and $BS^1 = \mathbb{C}P^\infty$ are infinite projective spaces, $B\mathbb{Z}/p = S^\infty/(\mathbb{Z}/p)$ is an infinite lens space and

$$BO(n) = \{V \subset \mathbb{R}^\infty : \dim V = n\}$$

is the infinite *Grassmannian manifold of n-planes* in $\mathbb{R}^\infty$.

For a G-space X we write

$$EG \underset{G}{\times} X := (EG \times X)/G$$

where G acts diagonally on $EG \times X$. For a G-pair (X, X') we can now define the *Borel cohomology* by

$$H_G^*(X, X'; R) := H^*(EG \underset{G}{\times} X, EG \underset{G}{\times} X'; R).$$

At first it seems more natural to take the cohomology of the orbit space, that is $H^*(X/G; R)$, as the equivariant cohomology of X. This is possible but difficult to deal with because the projection $X \to X/G$ is not a bundle in general. When one replaces X by $EG \times X$ one observes two effects: $EG \times X$ is homotopy equivalent to X because EG is contractible; and $EG \times X \to EG \underset{G}{\times} X$ is a bundle with fibre G because the action of G on $EG \times X$ is free. Thus without changing X too much one obtains a more manageable object. If the action of G on X was already free then $EG \underset{G}{\times} X$ is homotopy equivalent to X/G, hence $H_G^*(X; R) \cong H^*(X/G; R)$.

Clearly, H_G^* is multiplicative because Alexander-Spanier cohomology has a cup-product ([Sp], p. 315). Similarly, H_G^* is continuous since this is true for H^* ([Sp], Theorem 6.6.2). Observe that $EG \underset{G}{\times} X$ is metrizable if X is. Here it is important that G is compact. The continuity property even holds when X is only paracompact ($EG \underset{G}{\times} X$ is then also paracompact because EG is σ-compact) but we shall not leave the realm of metrizable spaces.

If H is a closed subgroup of G then the inclusion $EH \simeq EG \hookrightarrow EG \times (G/H)$ induces a homotopy equivalence of the orbit spaces $BH = EH/H \simeq EG/H \cong EG \underset{G}{\times} (G/H)$. This implies $H_G^*(G/H) \cong H^*(BH)$, in particular $H_G^*(\text{pt}) \cong H^*(BG)$. Moreover, the inclusion $H \hookrightarrow G$ induces a map $BH \simeq EG/H \to EG/G = BG$. hence a homomorphism $H^*(BG) \longrightarrow H^*(BH)$. This corresponds to the map $H_G^*(\text{pt}) \longrightarrow H_G^*(G/H)$ induced by the G-map $G/H \to \text{pt}$. Finally, if $G = G_1 \times G_2$ then $BG \simeq BG_1 \times BG_2$. Thus one can use the Künneth theorem to compute $H_G^*(\text{pt}) \cong H^*(BG_1 \times BG_2)$.

We shall particularly refer to the following special cases.

$G = \mathbb{Z}/2$: We take $R = \mathbb{F}_2$ the field of two elements. $H_G^*(\mathrm{pt}) = H^*(\mathbb{RP}^\infty; \mathbb{F}_2) \cong \mathbb{F}_2[w]$ is a polynomial ring with one generator $w \in H_G^1(\mathrm{pt})$.

$G = \mathbb{Z}/p$, p an odd prime: We take $R = \mathbb{F}_p$ the field of p elements. The coefficient ring $H_G^*(\mathrm{pt}) \cong \mathbb{F}_p[c] \otimes E[w]$ is the tensor product of a polynomial ring with one generator $c \in H_G^2(\mathrm{pt})$ and an exterior algebra with one generator $w \in H_G^1(\mathrm{pt})$. In particular $w^2 = 0$.

$G = S^1$: We take rational coefficients $R = \mathbb{Q}$. $H_G^*(\mathrm{pt}) = H^*(\mathbb{CP}^\infty; \mathbb{Q}) \cong \mathbb{Q}[c]$ is a polynomial ring with one generator $c \in H_G^2(\mathrm{pt})$.

For tori $(\mathbb{Z}/p)^k$ respectively $(S^1)^k$ we take the same coefficients as for $\mathbb{Z}/p$ respectively S^1. The coefficient ring $H_G^*(\mathrm{pt})$ is then obtained via the Künneth theorem. An important example for us is the case $G = (S^1)^k$ where $H_G^*(\mathrm{pt}) \cong \mathbb{Q}[c_1, \ldots, c_k]$. Observe that in the examples considered so far the coefficient rings are *noetherian*, that is they are commutative and every ideal is finitely generated.

In the definition of Borel cohomology one can replace Alexander-Spanier cohomology by other (generalized) cohomology theories. For example, in §5.4 we use Borel stable cohomotopy theory in order to deal with finite p-groups. Occasionally we shall also work with equivariant cohomology theories which cannot be obtained via the Borel construction. In §5.3, for instance, where G is a cyclic p-group, we apply equivariant K-theory. The proper choice of h^* depends very much on the group G, as we shall see in Chapter 5.

4.3 The length

As a motivation for the definition of the length we first discuss the notion of cup-length in the non-equivariant context. Given a non-empty connected topological space X the *cup-length* of X (with respect to $H^* = H^*(-; R)$) is the smallest natural number $k \geq 1$ such that for all $\omega_1, \ldots, \omega_k \in \widetilde{H}^*(X)$ the cup-product $\omega_1 \smile \ldots \smile \omega_k$ is zero. The cup-length of the empty set is 0. Recall that

$$\widetilde{H}^*(X) = \mathrm{coker}\big(H^*(\mathrm{pt}) \to H^*(X)\big) \cong \ker\big(\beta^*: H^*(X) \to H^*(\mathrm{pt})\big)$$

for any map $\beta: \mathrm{pt} \to X$. If X is not connected then the cup-length of X is the smallest k such that there exist maps $\beta_i: \mathrm{pt} \to X$, $i = 1, \ldots, k$, with the property that for all $\omega_i \in \ker\big(\beta_i^*: H^*(X) \to H^*(\mathrm{pt})\big)$ the product $\omega_1 \smile \ldots \smile \omega_k$ is zero. This differs slightly from the usual definition where one requires $\omega_i \in H^*(X)$ with $* > 0$. Our definition has the advantage that the cup-length is additive (as is the category):

$$\text{cup-length}(X_1 \sqcup \ldots \sqcup X_n) = \sum_{i=1}^n \text{cup-length}(X_i).$$

This is an immediate consequence of the ring isomorphism

$$H^*(X_1 \sqcup \ldots \sqcup X_n) \cong \prod_{i=1}^n H^*(X_i).$$

It is easy to see that this is a lower bound for the Lusternik-Schnirelmann category of X. A relative version of this, i.e. the cup-length of a pair (X, X') of topological spaces $X' \subset X$, is the smallest k such that there exist maps $\beta_i \colon \mathrm{pt} \to X$, $i = 1, \ldots, k$, with the property that for all $\gamma \in H^*(X, X')$ and all $\omega_i \in \ker \beta_i^*$ the product $\omega_1 \smile \ldots \smile \omega_k \smile \gamma$ is zero in $H^*(X, X')$.

The proper equivariant version goes as follows. Fix a set $\mathcal{A}$ of G-spaces and a multiplicative equivariant cohomology theory h^*. The $(\mathcal{A}, h^*)$-*cup-length* of a pair (X, X') of G-spaces is the smallest k such that there exist "$\mathcal{A}$-points" $A_1, \ldots, A_k \in \mathcal{A}$ and G-maps $\beta_i \colon A_i \to X$, $i = 1, \ldots, k$, with the property that for all $\gamma \in h^*(X, X')$ and all $\omega_i \in \ker \beta_i^*$ the product $\omega_1 \smile \ldots \smile \omega_k \smile \gamma$ is zero in $h^*(X, X')$.

The $(\mathcal{A}, h^*)$-cup-length of a G-map $f \colon (X, X') \to (Y, Y')$ can be defined analogously (compare the definition of $\mathcal{A}$-cat(f) in 2.6). It is the smallest integer $k \geq 0$ such that there exist $A_1, \ldots, A_k \in \mathcal{A}$ and G-maps $\beta_i \colon A_i \to Y$ satisfying

$$f^*(\omega_1 \smile \ldots \smile \omega_k \smile \gamma) = 0 \in h^*(X, X')$$

for all $\omega_i \in \ker(\beta_i^*)$ and all $\gamma \in h^*(Y, Y')$. If $k = 0$ this means $\mathrm{im}(f^*) = 0$. This serves as a lower bound for $\mathcal{A}$-cat(f). Namely, if $k = \mathcal{A}$-cat(f) then by definition there exists a numerable covering $\{X_0, X_1, \ldots, X_k\}$ of X having certain properties (see Definition 2.6). In particular, there exist G-maps $\alpha_i \colon X_i \to A_i$ and $\beta_i \colon A_i \to Y$ such that the restriction $f|X_i$ is G-homotopic to $\beta_i \circ \alpha_i$. Given $\omega_i \in \ker(\beta_i^*)$ the image $f^*(\omega_i) \in h^*(X)$ restricts therefore to 0 in $h^*(X_i)$. Hence, the exact sequence of the pair (X, X_i) shows that $f^*(\omega_i)$ has a pre-image in $h^*(X, X_i)$. Similarly, for any $\gamma \in h^*(Y, Y')$ the image $f^*(\gamma) \in h^*(X, X')$ restricts to zero in $h^*(X_0, X')$ by part (i) of Definition 2.6. By the exact sequence of the triple (X, X_0, X') we conclude that $f^*(\gamma)$ has a pre-image in $h^*(X, X_0)$. Therefore the product

$$f^*(\omega_1 \smile \ldots \smile \omega_k \smile \gamma) = f^*(\omega_1) \smile \ldots \smile f^*(\omega_k) \smile f^*(\gamma)$$

has a pre-image in $h^*(X, X_0 \cup X_1 \cup \ldots \cup X_k) = 0$, so it must be zero. (The product is defined because the covering is numerable.) This shows that

$$(\mathcal{A}, h^*)\text{-cup-length}(f) \leq k = \mathcal{A}\text{-cat}(f)$$

as claimed.

However, this equivariant cup-length does not have all the properties of the category, for example the continuity property and the deformation invariance as stated in Proposition 2.11 do not hold in general. And for the applications in chapters 6 to 9 we need even more properties than $\mathcal{A}$-cat or $\mathcal{A}$-genus satisfy. These considerations lead to the following modified definition. Recall that for a multiplicative equivariant cohomology theory h^* and a G-pair (X, X') the cohomology $h^*(X, X')$ is a module over the coefficient ring $R := h^*(\mathrm{pt})$ via the natural map $p_X \colon X \to \mathrm{pt}$. We write $\omega \cdot \gamma = p_X^*(\omega) \smile \gamma$ and $\omega_1 \cdot \omega_2 = \omega_1 \smile \omega_2$ for $\gamma \in h^*(X, X')$ and $\omega_1, \omega_2 \in R$.

4.1 Definition:

Fix a set $\mathcal{A}$ of G-spaces, a multiplicative equivariant cohomology theory h^* and an ideal I of the coefficient ring $R = h^*(\mathrm{pt})$. The $(\mathcal{A}, h^*, I)$-*length* of a G-map $f: (X, X') \to (Y, Y')$ is the smallest integer $k \geq 0$ such that there exist $A_1, \ldots, A_k \in \mathcal{A}$ with the following property:

For all $\gamma \in h^*(Y, Y')$ and all $\omega_i \in I \cap \ker(R \to h^*(A_i))$, $i = 1, \ldots, k$, the product $f^*(\omega_1 \cdot \ldots \cdot \omega_k \cdot \gamma) = \omega_1 \cdot \ldots \cdot \omega_k \cdot f^*(\gamma)$ is zero in $h^*(X, X')$.

If $k = 0$ this means that $f^* = 0$. If no such k exists we set $(\mathcal{A}, h^*, I)$-$\mathrm{length}(f) := \infty$. We write $(\mathcal{A}, h^*, I)$-$\mathrm{length}(X, X')$ for $(\mathcal{A}, h^*, I)$-$\mathrm{length}(\mathrm{id}_{(X, X')})$.

Observe that we do not use the structure of $h^*(X, X')$ as a module over $h^*(X)$ but only the reduced structure as an R-module. The introduction of the ideal $I \subset R$ is a refinement which allows a more detailed investigation in certain cases (e.g. for $G = \mathbb{Z}/p$). A purely algebraic version of the length can be defined as follows. Fix a ring R and a set $\mathcal{I}$ of ideals of R. Then the $(R, \mathcal{I})$-*length* of a (left) R-module M is the smallest integer $k \geq 0$ such that there exist $I_1, \ldots, I_k \in \mathcal{I}$ with the following property:

The product ideal $I_1 \cdot \ldots \cdot I_k$ annihilates M: $I_1 \cdot \ldots \cdot I_k \cdot M = 0$.

Given a triple $(\mathcal{A}, h^*, I)$ as in Definition 4.1 we set $\mathcal{I} = \{\ker(I \to h^*(A)): A \in \mathcal{A}\}$. Then $(\mathcal{A}, h^*, I)$-$\mathrm{length}(f) = (R, \mathcal{I})$-$\mathrm{length}(\mathrm{im}\ f^*)$ where $R = h^*(\mathrm{pt})$, of course. Many of the properties of the $(\mathcal{A}, h^*, I)$-length can be reformulated in this general setting, at least if R is a (graded) commutative ring and has a unit.

Whereas the $(\mathcal{A}, h^*)$-cup-length restricts to the h^*-cup-length if G is the trivial group and $\mathcal{A} = \{\mathrm{pt}\}$, the $(\mathcal{A}, h^*, I)$-length is trivial in this case because $\ker(R \to h^*(\mathrm{pt})) = 0$. In other words, the length is not a generalization of the cup-length. We shall see in the next sections that it has properties which are very useful in critical point theory and which do not hold for the cup-length.

It is also clear that the $(\mathcal{A}, h^*, I)$-length is less than or equal to the $(\mathcal{A}, h^*)$-cup-length as defined before 4.1. Namely, if $\omega_i \in \ker(I \to h^*(A_i))$ then $p_Y^*(\omega_i) \in \ker(\beta_i^*: h^*(Y) \to h^*(A_i))$ for any G-map $\beta_i: A_i \to Y$. Thus

$$f^*(\omega_1 \cdot \ldots \cdot \omega_k \cdot \gamma) = f^*\left(p_Y^*(\omega_1) \smile \ldots \smile p_Y^*(\omega_k) \smile \gamma\right) = 0$$

if $k = (\mathcal{A}, h^*)$-cup-length(f). This gives the next result.

4.2 Proposition:

For any triple $(\mathcal{A}, h^*, I)$ as in Definition 4.1 and any G-map $f: (X, X') \to (Y, Y')$

$$(\mathcal{A}, h^*, I)\text{-length}(f) \leq \mathcal{A}\text{-cat}(f).$$

□

One can also show that the $(\mathcal{A}, h^*, I)$-length of a G-space X is a lower bound of $\mathcal{A}$-genus(X); see Corollary 4.9a) below. Before studying the length in detail we look at some important examples. If $(\mathcal{A}, h^*, I)$ is understood we write ℓ for $(\mathcal{A}, h^*, I)$-length.

4.3 Example:

For the group $G = S^1$ we always use Borel cohomology (cf. §4.2) with rational coefficients: $h^* = H_G^*(-; \mathbb{Q})$. The coefficient ring R is isomorphic to the polynomial ring $\mathbb{Q}[c]$ on one generator $c \in h^2(\mathrm{pt})$. Setting

$$\mathcal{A} = \{G/H\colon H \subsetneqq G \text{ a closed subgroup}\}$$

and $I = R$, the length of a G-map f is

$$\ell(f) = \min\left\{k \geq 0\colon c^k \text{ annihilates } \mathrm{im}(f^*)\right\}.$$

Here we used the fact that $EG \underset{G}{\times} (G/H) \simeq BH$, so

$$h^*(G/H) \cong H^*(BH; \mathbb{Q}) \cong H^*(EH; \mathbb{Q})^H \cong H^0(EH; \mathbb{Q}) \cong \mathbb{Q}$$

for a finite subgroup H of G. Therefore $\ker\left(R \to h^*(G/H)\right)$ is the ideal $(c) \subset R$. The second isomorphism above is a consequence of a result of Grothendieck [Gro], p. 202; cf. also [Bor], Corollary III.2.3. For a G-space X we get

$$\ell(X) = \min\left\{k \geq 0\colon c^k \cdot 1_X = 0 \in h^*(X)\right\}.$$

If $G = (S^1)^r$ is a torus we set $\mathcal{A} = \{G/H\colon H \subsetneqq G \text{ a closed subgroup}\}$. Again using rational coefficients the Künneth theorem yields $R \cong \bigotimes_{i=1}^r \mathbb{Q}[c_i] \cong \mathbb{Q}[c_1, \ldots, c_r]$ where $c_i \in h^2(\mathrm{pt})$ generates the image of $p_i^*\colon H^2(BS^1; \mathbb{Q}) \longrightarrow H^2(BG; \mathbb{Q}) = h^2(\mathrm{pt})$. Here $p_i\colon BG \to BS^1$ is induced from the projection $G \to S^1$ onto the i-th factor. The length of a G-map f is more complicated:

$$\ell(f) = \min\left\{k \geq 0\colon \exists (n_{ij}) \in \mathbb{Z}^{k \times r} \text{ such that } \prod_{i=1}^k \left(\sum_{j=1}^r n_{ij} c_j\right) \in R \text{ is non-zero} \right.$$
$$\left. \text{and annihilates } \mathrm{im}(f^*)\right\}.$$

4.4 Example:

For the group $G = \mathbb{Z}/2$ we shall always use Borel cohomology (see §4.2) with coefficients in the field $\mathbb{F}_2$ of two elements: $h^* = H_G^*(-; \mathbb{F}_2)$ where H^* is Alexander-Spanier cohomology. The coefficient ring $R = h^*(\mathrm{pt})$ is isomorphic to the polynomial ring $\mathbb{F}_2[w]$ on one generator $w \in h^1(\mathrm{pt})$. Setting $\mathcal{A} = \{G\}$ and $I = R$ the length of a G-map f is

$$\ell(f) = \min\{k \geq 0\colon w^k \text{ annihilates } \mathrm{im}(f^*)\}.$$

Here we used the fact that $h^*(G) \cong H^*(\mathrm{pt}; \mathbb{F}_2)$, so $\ker\left(h^*(\mathrm{pt}) \to h^*(G)\right)$ is the ideal $(w) \subset R$. For a G-space X we obtain

$$\ell(X) = \min\{k \geq 0\colon w^k \cdot 1_X = 0 \in h^*(X)\}.$$

If $G = (\mathbb{Z}/2)^r$ is a 2-torus we use again $\mathbb{F}_2$-coefficients and obtain $R \cong \mathbb{F}_2[w_1, \ldots, w_r]$ with $w_i \in h^1(\text{pt})$ generating the image of $p_i^*: H^1(B\mathbb{Z}/2; \mathbb{F}_2) \longrightarrow H^1(BG; \mathbb{F}_2)$. The length of a G-map can be described as in 4.3 with c_i replaced by w_i.

4.5 Example:

For the group $G = \mathbb{Z}/p$, $p > 2$ a prime, we use $h^* = H_G^*(-; \mathbb{F}_p)$. The coefficient ring R is isomorphic to the tensor product $\mathbb{F}_p[c] \otimes E[w]$ of the polynomial ring $\mathbb{F}_p[c]$ on one generator $c \in h^2(\text{pt})$ and the exterior $\mathbb{F}_p$-algebra $E[w]$ on one generator $w \in h^1(\text{pt})$. This just means that $w^2 = 0$ so that $h^*(\text{pt}) \cong \mathbb{F}_p[c, w]/(w^2)$. Setting $\mathcal{A} = \{G\}$ and $I_0 = (c) \subset R$, $I_1 = R$, we define two versions of the length: $\ell_j = (\mathcal{A}, h^*, I_j)$-length for $j = 0, 1$. Then we get for a G-map f

$$\ell_0(f) = \min\{k \geq 0: c^k \text{ annihilates im } f^*\},$$

and for a G-space X

$$\ell_0(X) = \min\{k \geq 0: c^k \cdot 1_X = 0 \in h^*(X)\}.$$

And $\ell_1(f)$ can be described as follows:

$$\ell_1(f) = \begin{cases} \ell_0(f) =: k & \text{if } c^{k-1}w \text{ annihilates im } f^*, \\ \ell_0(f) + 1 & \text{if not.} \end{cases}$$

If G is a p-torus we proceed as in the examples 4.3 and 4.4 above and obtain

$$R \cong \mathbb{F}_p[c_1, \ldots, c_r] \otimes E[w_1, \ldots, w_r].$$

There are various choices for the ideal I. We only mention one: Setting $I_0 = (c_1, \ldots, c_r)$ the length of a G-map can be described as in Example 4.3.

4.6 Remark:

For $G = \mathbb{Z}/2$ the length of a G-space X as defined in Example 4.4 has first been introduced by Yang in [Y1] as *the cohomological index* of X. Yang was interested in generalizations of the Borsuk-Ulam theorem. Fadell and Rabinowitz [FaR1] were the first who used this index in critical point theory. So far, no proof of Fadell's and Rabinowitz' result or of later generalizations (as in [FaR2], [BC1]; see also §7.5) is known which works only with the category or the genus. This is due to the fact that certain properties of the length are not true for category or genus. In [FaR2] Fadell and Rabinowitz also introduced the length for $G = S^1$ which is an important case from the point of view of applications. In a series of joint papers with Husseini (cf. [FaHR, FaH1,2]) they developed more general cohomological index theories. Their most general version is the following one. Given G and a multiplicative equivariant cohomology theory h^* the h^*-index of a G-pair (X, X') is the ideal of $R = h^*(\text{pt})$ which annihilates $h^*(X, X')$. In particular, h^*-index$(X) = \ker\left(p_X^*: R \to h^*(X)\right)$. So this index is ideal-valued. They develop the theory only for $h^* = H_G^*(-; \mathbb{F})$. The numerical cohomological index of Fadell, Husseini and Rabinowitz is the rank of $H_G^*(\text{pt})/H_G^*$-index(X) as a module over $\mathbb{F}$. The length (in a slightly simpler version than in 4.1) has been introduced and applied in a series of papers of Clapp, Puppe and the author ([ClP2], [BCP], [BC1]). In [ClP2] and [BCP] the length is merely used as a lower bound for the category. In [BCP] equivariant stable cohomotopy has

been used for the first time instead of Borel cohomology. And in [BC1] the result
of [FaR2] has been generalized using properties of the length which do not hold for
the category.

Let us compare the length with the cohomological index of Fadell, Husseini
and Rabinowitz. First of all, it is important to work not only with Borel cohomology but also with other equivariant cohomology theories depending on the group
G; see for example chapter 5. Fixing h^*, the basic ingredient of all cohomological indexes of a G-space X is the kernel of the homomorphism $h^*(\text{pt}) \to h^*(X)$.
Modeled on the definition of the $\mathcal{A}$-category or the $\mathcal{A}$-genus the $(\mathcal{A}, h^*, I)$-length
captures precisely those properties needed in critical point theory. The choice of
$\mathcal{A}$ reflects the orbit structure of the set of critical points one looks for; usually
$\mathcal{A} \subset \{G/H : H \subsetneqq G \text{ a closed subgroup}\}$. The more one knows a priori about this
orbit structure the smaller one can choose $\mathcal{A}$. The choice of I makes the length in
principle as strong as the ideal-valued index. For instance, in order to prove the
non-existence of a G-map $X \to Y$ with the help of the h^*-index one has to show
that h^*-index$(Y) \not\subset h^*$-index(X). But if this is true then there also exists an ideal
I such that $(\mathcal{A}, h^*, I)$-length$(Y) < (\mathcal{A}, h^*, I)$-length$(X)$; here $\mathcal{A}$ contains (up to
homeomorphism) all orbits in $X \cup Y$. Choose an element ω of h^*-index(Y) which
is not contained in h^*-index(X) and set $I = (\omega) \subset R$. Then it is clear that

$$(\mathcal{A}, h^*, I)\text{-length}(Y) = 1 < (\mathcal{A}, h^*, I)\text{-length}(X).$$

This also implies that no G-map $X \to Y$ exists due to the monotonicity of the
length (see Theorem 4.7 below).

From this point of view the ideal-valued h^*-index is replaced by a family of
numerical indexes, one for each pair $(\mathcal{A}, I)$. The proper choice of $(\mathcal{A}, I)$ depends on
the situation. One certainly looses more information when one passes to the rank
of $H_G^*(\text{pt})/H_G^*$-index. Already for tori and p-tori this numerical index is too weak.
In fact, Fadell and Husseini introduced the ideal-valued H_G^*-index in order to deal
with the 2-torus $G = (\mathbb{Z}/2)^k$. As we shall see in §5.2 there is a natural choice of
$(\mathcal{A}, I)$ for these groups so that one only has to work with one numerical index.

A variation of the length corresponding to cat_M is useful when one is only
interested in subsets of a fixed G-space M, for example when one does critical point
theory for functionals $M \to \mathbb{R}$. Then one can replace the coefficient ring $R = h^*(\text{pt})$
in Definition 4.1 by the cohomology ring of M. If $X' \subset X \subset M$ are G-subsets of
M then $h^*(X, X')$ is a module over $h^*(M)$, not only over $h^*(\text{pt})$. The properties of
the length as in §4.4 can be generalized easily, but we shall not develop this point
of view here. An ideal-valued cohomological index theory using this idea is due to
Komiya [Ko1].

4.4 Properties of the length

Now we begin to investigate the length more closely. We fix a triple $(\mathcal{A}, h^*, I)$ and write ℓ for $(\mathcal{A}, h^*, I)$-length.

4.7 Theorem:
The $(\mathcal{A}, h^*, I)$-length ℓ has the following properties.

Monotonicity: *If there exists a G-map $f\colon X \to Y$ then $\ell(X) \le \ell(Y)$.*

Subadditivity: *If X' and X'' are invariant subspaces of X whose interiors cover X then*

$$\ell(X' \cup X'') \le \ell(X') + \ell(X'').$$

Excision: *If X' and X'' are invariant subspaces of X whose interiors cover X then*

$$\ell(X'', X' \cap X'') = \ell(X, X').$$

Intersection: *Let X' and X'' be invariant subspaces of X whose interiors cover X and such that the inclusion $X' \hookrightarrow X$ induces a monomorphism $h^*(X) \to h^*(X')$, for example X can be deformed into X' (i.e. the inclusion has a right homotopy inverse). Then*

$$\ell(X'') = \ell(X' \cap X'').$$

Triangle inequality: *Given a triple $X'' \subset X' \subset X$ of G-spaces, then the sum of any two of $\ell(X, X')$, $\ell(X, X'')$, $\ell(X', X'')$ is greater than or equal to the third; for example*

$$\ell(X, X') + \ell(X', X'') \ge \ell(X, X'').$$

Normalization: *For any $A \in \mathcal{A}$*

$$\ell(A \sqcup \ldots \sqcup A) = 1.$$

Continuity: *If h^* is continuous and I is noetherian (as R-module) then any locally closed G-subset X' of a metrizable G-space X has an open invariant neighborhood U with $\ell(U) = \ell(X')$.*

Recall that I *noetherian* implies that every R-module contained in I is finitely generated. This is automatically the case if R is noetherian as in the examples mentioned so far. The proof of Theorem 4.7 is postponed to the end of this section. We continue to discuss the length. The next result is a simple consequence of the monotonicity property.

4.8 Corollary:
For any G-space X: $\ell(X) \le \ell(\mathrm{pt})$, and equality holds if $X^G \ne \emptyset$. □

One checks easily that $\ell(\mathrm{pt}) = \infty$ in the examples 4.3 to 4.5. This is not true in general. For instance, if pt $\in \mathcal{A}$ then $\ell(\mathrm{pt}) = 1$. In this case we cannot obtain any interesting information at all using the length. For the applications we have in mind, the bigger $\ell(\mathrm{pt})$, the better. In fact, we usually apply ℓ only if $\ell(\mathrm{pt}) = \infty$. From the definition it follows that

$$\mathcal{A}' \subset \mathcal{A} \Rightarrow (\mathcal{A}, h^*, I)\text{-length} \leq (\mathcal{A}', h^*, I)\text{-length}.$$

Therefore we want to make $\mathcal{A}$ as small as possible. The choice of $\mathcal{A}$ is often forced by the application as, for instance, in the critical point theorem 2.19 where $\mathcal{A}$ has to contain all critical G-orbits (up to G-homeomorphism) in which one is interested. So one always has to make sure that no critical point is fixed by G if one wants to work with the length. But even if $\mathcal{A}$ is a subset of $\{G/H \colon H \subsetneqq G$ a closed subgroup$\}$ one has to choose h^* (and I) carefully in order to assure $\ell(\mathrm{pt}) = \infty$ — if this is possible at all. We shall address this problem in §5.5.

4.9 Corollary:

a) $\ell(X) \leq \mathcal{A}\text{-genus}(X)$ *for every G-space X.*

b) $\ell(X) < \infty$ *if X is a compact G-space such that each orbit of X can be mapped equivariantly into some element of $\mathcal{A}$.*

Proof:

a) If $X \to A_1 * \ldots * A_k$ is a G-map as in the definition of $k = \mathcal{A}\text{-genus}(X)$ then

$$\ell(X) \leq \ell(A_1 * \ldots * A_k) \leq \sum_{i=1}^{k} \ell(A_i) = k = \mathcal{A}\text{-genus}(X).$$

Here we used the monotonicity and the normalization property of ℓ as well as Corollary 4.10 below.

b) This follows from a) and Proposition 2.16a).   □

Recall that the join of two G-spaces X and Y is the G-space

$$X * Y := \{[t, x, 1 - t, y] \colon 0 \leq t \leq 1, x \in X, y \in Y\}$$

with the identifications $[0, x, 1, y] = [0, x', 1, y]$ and $[1, x, 0, y] = [1, x, 0, y']$. Identifying $x \in X$ with $[1, x, 0, y]$ and $y \in Y$ with $[0, x, 1, y]$ we can consider X and Y as closed G-subspaces of $X * Y$. Now $\ell(X * Y - X) \leq \ell(Y)$ and $\ell(X * Y - Y) \leq \ell(X)$ since the obvious projections $X * Y - X \to Y$ and $X * Y - Y \to X$ are equivariant. Using the subadditivity of ℓ this proves

4.10 Corollary:

For any two G-spaces X and Y we have $\ell(X * Y) \leq \ell(X) + \ell(Y)$.   □

The following piercing property of ℓ was already known to Yang for his $\mathbb{Z}/2$-index (see [Y3], Proof of Theorem (4.1)). It was rediscovered and emphasized by Fadell and Rabinowitz [FaR1,2] because it played a crucial role in the proof of their bifurcation results and because it does not hold for the category.

4.11 Corollary:
Piercing property: *Let X_0 and X_1 be invariant subsets of X whose interiors cover X. Consider a G-space Y and a G-deformation $\varphi_t \colon Y \to X$, $t \in [0,1]$, satisfying $\varphi_i(Y) \subset X_i$ for $i = 0,1$. Then*

$$\ell\bigl(\operatorname{im}(\varphi) \cap X_0 \cap X_1\bigr) \geq \ell(Y).$$

If φ is an imbedding equality holds.

Proof:
Set $Z_i := \varphi^{-1}(X_i) \subset Z := Y \times [0,1]$, $i = 0,1$. By the monotonicity of ℓ it suffices to show that $\ell(Z_0 \cap Z_1) \geq \ell(Y)$ because φ induces a G-map $Z_0 \cap Z_1 \to \operatorname{im}(\varphi) \cap X_0 \cap X_1$. Now $Y \times \{i\} \subset Z_i$, $i = 0,1$, hence Z_i contains a deformation retract of $Y \times [0,1]$. Applying the intersection property to the triad $(Z; Z_0, Z_1)$ and once more the monotonicity of ℓ one gets

$$\ell(Z_0 \cap Z_1) = \ell(Z_1) = \ell(Y).$$

If φ is an imbedding then $Z_0 \cap Z_1$ is G-homeomorphic to $\operatorname{im}(\varphi) \cap X_0 \cap X_1$ which implies equality of the lengths. □

4.12 Remark:
Suppose h^* is continuous. Then the subadditivity property also holds if $X = X' \cup X''$ is metrizable and the invariant subspaces X' and X'' are locally closed. It is not necessary that X is covered by the interiors of X' and X''. Similarly, the intersection and, hence, the piercing property hold if $X = X' \cup X''$ is metrizable and X', X'' are closed G-subsets of X.

It remains to prove Theorem 4.7.

Proof of Monotonicity:
This is a trivial consequence of the fact that any G-map $f \colon X \to Y$ induces a homomorphism $h^*(Y) \to h^*(X)$ of R-algebras with unit. And if $\omega \in R$ annihilates $1 \in h^*(X)$ then it annihilates $h^*(X)$. □

Proof of Subadditivity:
If $k = \ell(X')$ and $m = \ell(X'')$ choose $A_1, \ldots, A_k \in \mathcal{A}$ and $A_{k+1}, \ldots, A_{k+m} \in \mathcal{A}$ as in the definition of ℓ. Given $\omega_i \in I \cap \ker\bigl(R \to h^*(A_i)\bigr)$ the product $\omega_1 \cdot \ldots \cdot \omega_k \cdot 1_X \in h^*(X)$ maps to zero in $h^*(X')$, hence it comes from $h^*(X, X')$ as one sees from the long exact sequence of (X, X'). Similarly, the product $\omega_{k+1} \cdot \ldots \cdot \omega_{k+m} \cdot 1_X \in h^*(X)$ comes from $h^*(X, X'')$. Then $\omega_1 \cdot \ldots \cdot \omega_{k+m} \cdot 1_X$ comes from $h^*(X, X' \cup X'') = 0$. This proves $\ell(X) \leq k + m = \ell(X') + \ell(X'')$. Concerning Remark 4.12, if h^* is continuous then the interior product is also defined when X', X'' are locally closed subsets of the metrizable space X. □

Proof of Excision:
This is trivial because $h^*(X'', X' \cap X'') \cong h^*(X, X')$ as R-modules. □

Proof of Intersection:

We use the following part of the Mayer-Vietoris sequence of $(X; X', X'')$:

$$\ldots \longrightarrow h^*(X) \longrightarrow h^*(X') \oplus h^*(X'') \longrightarrow h^*(X' \cap X'') \longrightarrow \ldots$$

Consider an $\alpha \in h^*(X'')$ with $\alpha|(X' \cap X'') = 0$. By exactness, $(0, \alpha)$ comes from $\beta \in h^*(X)$, so $\beta|X' = 0$. This implies $\beta = 0$ by our assumption on X', hence $\alpha = \beta|X'' = 0$. Thus $h^*(X'') \to h^*(X' \cap X'')$ is injective and therefore $\ell(X'') = \ell(X' \cap X'')$. Concerning 4.12, if h^* is continuous then the Mayer-Vietoris sequence exists also for metrizable X and closed G-subsets X', X'' of X. $\quad\square$

Proof of Triangle Inequality:

We show $\ell(X, X') + \ell(X', X'') \geq \ell(X, X'')$. Consider the following part of the long exact sequence of the triple (X, X', X''):

$$\ldots \longrightarrow h^*(X, X') \xrightarrow{i^*} h^*(X, X'') \xrightarrow{j^*} h^*(X', X'') \longrightarrow \ldots$$

If $k = \ell(X, X')$ and $m = \ell(X', X'')$ choose $A_1, \ldots, A_k \in \mathcal{A}$ and $A_{k+1}, \ldots, A_{k+m} \in \mathcal{A}$ according to the definition of ℓ. Let $\gamma \in h^*(X, X'')$ and $\omega_i \in \ker\big(R \to h^*(A_i)\big)$, $i = 1, \ldots, k + m$, be arbitrary. We show that $\omega_1 \cdot \ldots \cdot \omega_{k+m} \cdot \gamma = 0$ which implies $\ell(X, X'') \leq k + m$. Now

$$j^*(\omega_{k+1} \cdot \ldots \cdot \omega_{k+m} \cdot \gamma) = \omega_{k+1} \cdot \ldots \cdot \omega_{k+m} \cdot j^*(\gamma) = 0,$$

so there exists $\beta \in h^*(X, X')$ with $i^*(\beta) = \omega_{k+1} \cdot \ldots \cdot \omega_{k+m} \cdot \gamma$. Now $\omega_1 \cdot \ldots \cdot \omega_k \cdot \beta = 0$, hence

$$\omega_1 \cdot \ldots \cdot \omega_{k+m} \cdot \gamma = i^*(\omega_1 \cdot \ldots \cdot \omega_k \cdot \beta) = 0.$$

Here we only used that i^* and j^* are R-module homomorphisms. This is also true for the connecting homomorphism $h^*(X', X'') \to h^{*+1}(X, X')$. Thus the other two inequalities can be proved analogously. $\quad\square$

Proof of Normalization:

Using monotonicity it suffices to show $\ell(A) = 1$ for $A \in \mathcal{A}$. This is a trivial consequence of the definition of ℓ. $\quad\square$

Proof of Continuity:

First $\ell(X') \leq \ell(U)$ is true for any neighborhood U of X' because of monotonicity. Suppose $k = \ell(X') < \infty$. Take $A_1, \ldots, A_k$ as in Definition 4.1 and choose finitely many generators $\omega_{ij} \in I \cap \ker\big(R \to h^*(A_i)\big)$, $i = 1, \ldots, k$, $j = 1, \ldots, k_i$. Here we use that I is a noetherian R-module. For any k-tupel $J = (j_1, \ldots, j_k)$ the product $\omega_J := \omega_{1, j_1} \cdot \ldots \cdot \omega_{k, j_k}$ maps to zero in $h^*(X')$. Since $h^*(X') \cong \varinjlim h^*(U)$ there exists a neighborhood U_J of X' such that ω_J maps to zero in $h^*(U_J)$. Then $U := \bigcap_J U_J$ satisfies $\ell(U) \leq k$. $\quad\square$

4.5 More properties for special groups

In this section we prove some results for $\mathbb{Z}/p$ and S^1 which do not hold in general. We shall only use the lengths ℓ resp. ℓ_0, ℓ_1 defined in 4.3 to 4.5 for these groups. First we refine the subadditivity property for $\mathbb{Z}/p$ where p is an odd prime.

4.13 Proposition:

Consider the group $G = \mathbb{Z}/p$ with p an odd prime. Let X' and X'' be invariant subspaces of X whose interiors cover X. If $\ell_1(Z) = \ell_0(Z) + 1$ holds for some $Z \in \{X', X''\}$ then

$$\ell_1(X' \cup X'') \leq \ell_1(X') + \ell_1(X'') - 1.$$

Proof:

Without loss of generality $\ell_1(X') = \ell_0(X')+1$. If in addition $\ell_1(X'') = \ell_0(X'')+1$ then

$$\ell_1(X) \leq \ell_0(X) + 1 \leq \ell_0(X') + \ell_0(X'') + 1 = \ell_1(X') + \ell_1(X'') - 1.$$

If on the other hand $\ell_1(X'') = \ell_0(X'') =: m$ we set $k := \ell_0(X')$. In this case we have $c^k \cdot 1_{X'} = 0$ and $c^{m-1}w \cdot 1_{X''} = 0$, hence, as in the proof of the subadditivity property in §4.4 we see that $c^{k+m-1}w \cdot 1_X = 0 \in H_G^*(X)$ because it comes from $H_G^*(X, X' \cup X'') = 0$. Similarly, $c^{k+m} \cdot 1_X = 0 \in H_G^*(X)$. Thus

$$\ell_1(X) \leq k + m = \ell_0(X') + \ell_0(X'') = \ell_1(X') + \ell_1(X'') - 1.$$

$\square$

4.14 Remark:

Because of Proposition 4.13 one is tempted to define

$$\ell(X, X') := \begin{cases} \ell_0(X, X') + \ell_1(X, X') - 1 & \text{if } H_G^*(X, X') \neq 0, \\ 0 & \text{if } H_G^*(X, X') = 0, \end{cases}$$

for a $\mathbb{Z}/p$-pair (X, X'), p an odd prime. It is clear that this ℓ satisfies monotonicity, excision, intersection, normalization and continuity. Using the above proposition a simple argument yields the following version of the subadditivity property:

If X' and X'' are invariant subspaces of X whose interiors cover X and if $\ell_1(Z) = \ell_0(Z) + 1$ for some $Z \in \{X, X', X''\}$ then $\ell(X) \leq \ell(X') + \ell(X'')$.

If the additional assumption holds for $Z \in \{X', X''\}$ then we use 4.13. And if $\ell_1(Z) = \ell_0(Z)$ for $Z \in \{X', X''\}$ but $\ell_1(X) = \ell_0(X) + 1$ then

$$\ell(X) = 2\ell_1(X) - 2 \leq 2\big(\ell_1(X') + \ell_1(X'')\big) - 2 = \ell(X') + \ell(X'').$$

Similarly one can proof the triangle inequality:

Given a triple $X'' \subset X' \subset X$ of G-spaces such that $\ell_1(Z, Y) = \ell_0(Z, Y) + 1$ for some $(Z, Y) \in \{(X, X'), (X, X''), (X', X'')\}$ then the sum of any two of $\ell(X, X')$, $\ell(X, X'')$, $\ell(X', X'')$ is greater or equal to the third.

Since we do not have any counterexamples it is possible that ℓ does satisfy the sub-additivity property and the triangle inequality in general, i.e. without an additional assumption. Another interesting property of ℓ is the following one (cf. Remark 5.4b):

$$\ell(\underbrace{\mathbb{Z}/p * \ldots * \mathbb{Z}/p}_{k \text{ factors}}) = k.$$

One way to think of ℓ is that it does not just count the number of factors in a product $c^k \cdot w^l$ but also the dimensions of the factors:

$$\text{If } c^k \cdot w^l \cdot 1_X \neq 0 \in h^*(X) \text{ then } \ell(X) > 2k + l.$$

Thus c counts as 2 and w as 1. This formulation indicates similar generalizations for other groups. Many results which hold for $\mathbb{Z}/2$ can be generalized to $\mathbb{Z}/p$ using ℓ, at least if an additional assumption like those above holds. We shall mention such extensions to $\mathbb{Z}/p$ in the sequel.

4.15 Theorem:

Let $G = \mathbb{Z}/p$ or $G = S^1$ and let ℓ be any of the lengths defined in 4.3 to 4.5. Then $\ell(X \sqcup Y) = \max\{\ell(X), \ell(Y)\}$ for any two G-spaces X and Y.

Proof:

The result is an immediate consequence of the splitting

$$h^*(X \sqcup Y) \cong h^*(X) \oplus h^*(Y).$$

$\square$

4.16 Remark:

The result of Theorem 4.15 holds for $\ell = (\mathcal{A}, h^*, I)$-length if and only if it satisfies the following property:

Strong normalization: $\ell(A_1 \sqcup \ldots \sqcup A_k) = 1$ for any $A_1, \ldots, A_k \in \mathcal{A}$.

This property holds if and only if $\ell = (\widehat{\mathcal{A}}, h^*, I)$-length where $\widehat{\mathcal{A}}$ consists of all finite disjoint unions of elements of $\mathcal{A}$. An example for this, different from those defined in 4.3 to 4.5, is the following one. Consider the group $G = S^1 \times \Gamma$ where Γ is any finite group. Thus the connected component G_0 of G is S^1. Set

$$\mathcal{A} := \{G/H \colon H \subset G \text{ is a finite subgroup}\}$$

and

$$\mathcal{A}_0 := \{G_0/H \colon H \subset G_0 \text{ is a finite subgroup}\}.$$

Let $\ell_{G_0} := (\mathcal{A}_0, H^*_{G_0}(-; \mathbb{Q}), R_0)$-length be the usual length for $G_0 = S^1$ and set $\ell_G := (\mathcal{A}, H^*_G(-; \mathbb{Q}), R)$-length with $R := H^*_G(\mathrm{pt}; \mathbb{Q})$. We claim that $\ell_G(X) = \ell_{G_0}(X)$ for every G-space X. This holds in fact for every compact Lie group whose connected component G_0 is a torus. First observe that

$$R = H^*(BG; \mathbb{Q}) \cong H^*(BG_0; \mathbb{Q})^\Gamma = H^*(BG_0; \mathbb{Q}) = R_0 \cong \mathbb{Q}[c].$$

The second isomorphism is a consequence of [Gro], p. 202. We also used that Γ acts trivially on BG_0, hence on $H^*(BG_0;\mathbb{Q})$. Moreover, for any $G/H \in \mathcal{A}$ we have

$$\widetilde{H}_G^*(G/H;\mathbb{Q}) \cong \widetilde{H}^*(BH;\mathbb{Q}) \cong \widetilde{H}^*(EH;\mathbb{Q})^H \cong 0.$$

Thus $\ell_G(X)$ is the smallest integer $k \geq 0$ such that $c^k \cdot 1_X = 0 \in H_G^{2k}(X) \subset H_{G_0}^{2k}(X)$. This implies $\ell_G(X) = \ell_{G_0}(X)$ as claimed. In particular, ℓ_G satisfies the strong normalization property. We cannot conclude that $\ell_G = \ell_{G_0}$. But $\ell_G \leq \ell_{G_0}$ is a simple consequence of the fact that $H_G^*(X, X';\mathbb{Q}) \cong H_{G_u}^*(X, X';\mathbb{Q})^\Gamma \subset H_{G_0}^*(X, X';\mathbb{Q})$.

The next theorem is not so trivial. For $G = \mathbb{Z}/2$ it was already known to Yang (see [Y3], Theorem 2.3).

4.17 Theorem:

Let G be either the group $\mathbb{Z}/2$ or $S^1 \times \Gamma$ with Γ a finite group. Consider a finite-dimensional representation V of G. Then for any open G-subsets X, Y of the unit sphere SV we have:

$$\ell(X) + \ell(Y) \leq \ell(SV) + \ell(X \cap Y).$$

Here ℓ is as in Example 4.4 if $G = \mathbb{Z}/2$ or as in Remark 4.16 if $G = S^1 \times \Gamma$.

Proof:

First of all we may assume that $V^G = 0$ because otherwise $\ell(SV) = \infty$. If $G_0 = S^1$ we may even assume that $V^{G_0} = 0$ for the same reason. Since the result is known for $\mathbb{Z}/2$ and the proof is easier we shall only consider the case $G = S^1 \times \Gamma$. According to Remark 4.16 we have $\ell_G(Z) = \ell_{G_0}(Z)$ so that we may assume $G = G_0 = S^1$. Set $n := \frac{1}{2}\dim_{\mathbb{R}} V$. We use the fact that $H_G^*(SV) \cong \mathbb{Q}[c]/(c^n)$, so that $\ell(SV) = n$. This is a special case of Theorem 5.1. One also sees that $H_G^{2s}(SV)$ is generated by $c^s \cdot 1_{SV}$. Observe that the quotient map $q: EG \underset{G}{\times} SV \to SV/G$ induces an isomorphism $H^*(SV/G) \to H_G^*(SV)$ as a consequence of the Vietoris-Begle mapping theorem ([Sp], Theorem 6.9.15). Here we used that H^* is Alexander-Spanier cohomology with rational coefficients and

$$\widetilde{H}^*(q^{-1}(x)) \cong \widetilde{H}^*(EG \underset{G}{\times} Gx) \cong \widetilde{H}^*(BG_x) \cong 0.$$

The last isomorphism holds since $G_x \subset S^1$ is finite, so $\widetilde{H}^*(BG_x) \cong \widetilde{H}^*(EG_x)^{G_x} \cong 0$ according to [Gro]. If G_x is (up to conjugacy) independent of $x \in SV$, for example if G acts freely on SV, then $M := SV/G$ is a $(2n-2)$-dimensional orientable manifold. We shall first consider this case. (This part of the proof translates immediately to $G = \mathbb{Z}/2$.) Setting $k = \ell(X)$ and $m = \ell(Y)$ we have to show that $\ell(X \cap Y) \geq k + m - n =: r$. We may assume $r \geq 1$. In other words, we must show that the homomorphism

$$i^*: H_G^{2(r-1)}(SV) \cong H^{2(r-1)}(M) \longrightarrow H^{2(r-1)}\big((X \cap Y)/G\big) \cong H_G^{2(r-1)}(X \cap Y)$$

is not zero, that is, $i^*(c^{r-1} \cdot 1_{SV}) \neq 0$. By the universal coefficient theorem for cohomology ([Do], VI.7.8) this is equivalent to showing that $i_*: H_{2(r-1)}\big((X \cap Y)/G\big) \to$

$H_{2(r-1)}(M)$ is not zero. We know already that there exists $\xi \in H_{2(k-1)}(X/G)$ such that

$$i_*^X(\xi) \neq 0 \in H_{2(k-1)}(M) , \quad \text{where } i^X : X/G \hookrightarrow M,$$

because $k = \ell(X)$. Similarly there exists $\eta \in H_{2(m-1)}(Y/G)$ with $i_*^Y(\eta) \neq 0 \in H_{2(m-1)}(M)$. Let $x \in H^{2(n-k)}(M, M - X/G)$ and $y \in H^{2(n-m)}(M, M - Y/G)$ be the Poincaré-Lefschetz duals of ξ and η respectively ([Do], Proposition VIII.7.2). Then $j_X^*(x)$ is the dual of $i_*^X(\xi)$ where $j_X : M \to (M, M - X/G)$ denotes the inclusion (naturality of cap-products, [Do], VII.12.6). Therefore $j_X^*(x) \neq 0$, and since $H^{2(n-k)}(M)$ is generated by $c^{n-k} \cdot 1_M$ without loss of generality $j_X^*(x) = c^{n-k} \cdot 1_M$. Analogously, $j_Y^*(y) = c^{n-m} \cdot 1_M$. This implies

$$j_{X \cap Y}^*(x \smile y) = j_X^*(x) \smile j_Y^*(y) = c^{2n-k-m} \cdot 1_M = c^{n-r} \cdot 1_M \neq 0.$$

Applying duality once more we see that $i_* \neq 0$ as required. Observe that $x \smile y$ is dual to the intersection pairing $\xi \bullet \eta$ by the definition of $\bullet$ ([Do], VIII.13.1). So this part of the proof is more or less a translation of the proof of Yang [Y3] for $G = \mathbb{Z}/2$.

In the general case $M = SV/G$ is not a manifold but a rational cohomology manifold, that is, $H^*(M, M - x) \cong H^*(\mathbb{R}^{2n-2}, \mathbb{R}^{2n-2} - 0)$ for any $x \in M$. To see this consider a tubular neighborhood $U \cong G \underset{H}{\times} DW$ of the orbit $x = Gv \subset SV$. Here $W \cong \{w \in V : w \perp Gv\}$ is a $(2n-2)$-dimensional representation space of $H := G_v$. Then $x \in M$ has the neighborhood $U/G \cong DW/H$, hence

$$H^*(M, M - x) \cong H^*(DW/H, (DW - 0)/H) \cong \widetilde{H}^*(SW/H)$$
$$\cong \widetilde{H}^*(SW)^H \cong \widetilde{H}^*(SW).$$

For the second to last isomorphism it is again important that H is finite and that we use rational coefficients ([Gro]). And the last isomorphism is due to the fact that W is a complex representation of H. Therefore each $h \in H$ acts on W via a degree 1 map, hence it acts trivially on $\widetilde{H}^*(SW)$. Now the Poincaré-Lefschetz duality isomorphism also holds for $M = SV/G$ and rational cohomology. The proof as in [Do], VIII.7, can be easily adapted to the situation we are considering here. Therefore the above proof of the theorem in the special case where M is a manifold works in the general case, too. $\square$

Clearly, Theorem 4.17 also holds for closed G-subsets X, Y of SV since H_G^* is continuous. Easy examples show that Theorem 4.17 is false for tori and p-tori and the lengths defined in §4.3. The case $G = \mathbb{Z}/p$, p an odd prime, needs an extra treatment. Consider the lengths ℓ_0, ℓ_1 defined in Example 4.5. Theorem 4.17 is false for both ℓ_0 and ℓ_1. For instance, if $G \subset S^1$ acts on $V \cong \mathbb{C}$ via scalar multiplication let $A := Gv$ and $B := Gw$ be two different G-orbits of SV. Then $\ell_0(A) = 1 = \ell_0(B)$ and $\ell_0(A \cap B) = 0$. But Theorem 5.3 below tells us that $\ell_0(SV) = 1$, so 4.17 fails for ℓ_0. Next consider $W := V \oplus V$ and define $A := S(V \oplus 0)$, $B := S(0 \oplus V)$. Then $\ell_1(A) = 2 = \ell_1(B)$ and $\ell_1(SW) = 3$ according to 5.4a), but $\ell_1(A \cap B) = 0$, so 4.17 fails for ℓ_1, too. Observe that 4.17 does hold in these examples if one uses the "length" $\ell = \ell_0 + \ell_1 - 1$ as defined in Remark 4.14. We shall prove a weaker version of 4.17 for this ℓ which we need later.

4.18 Proposition:
  Let V be a finite-dimensional representation of $G = \mathbb{Z}/p$ where p is an odd prime and let ℓ be as in 4.14. Then for any open G-subsets X, Y of SV

$$\ell(X) + \ell(Y) \leq \ell(SV) + \ell(X \cap Y)$$

provided $\ell_1(X) = \ell_0(X) + 1$ (or $\ell_1(Y) = \ell_0(Y) + 1$).

Proof:
  We may assume that $V^G = 0$ so that SV is a free G-space. Then

$$R = H_G^*(\mathrm{pt}; \mathbb{F}_p) \cong \mathbb{F}_p[c] \otimes E[w]$$

and $H_G^*(SV; \mathbb{F}_p) \cong R/(c^n)$ where $n = \frac{1}{2} \dim V$; see §5.2. Therefore $\ell_0(SV) = n$ and $\ell_1(SV) = n + 1$, hence $\ell(SV) = 2n$. We consider two cases:

Case 1: $\ell_1(X) = \ell_0(X) + 1$ and $\ell_1(Y) = \ell_0(Y) + 1$;

Case 2: $\ell_1(X) = \ell_0(X) + 1$ and $\ell_1(Y) = \ell_0(Y)$.

  ad 1: Set $k := \ell_0(X)$ and $m := \ell_0(Y)$. As in the proof of Theorem 4.17 we shall show that $\ell(X \cap Y) \geq \ell(X) + \ell(Y) - \ell(SV) = 2k + 2m - 2n$. Observe that $M := SV/G$ is an orientable manifold with $\dim M = \dim SV = 2n - 1$ and $H^*(M) \cong H_G^*(SV)$; here we suppress the coefficients $\mathbb{F}_p$ from the notation. Since $c^{k-1}w \cdot 1_X \neq 0$ there exists $\xi \in H_{2k-1}(X/G)$ such that $i_*^X(\xi) \neq 0 \in H_{2k-1}(M)$. And there exists $\eta \in H_{2m-1}(Y/G)$ with $i_*^Y(\eta) \neq 0 \in H_{2m-1}(M)$. If $x \in H^{2n-2k}(M, M - X/G)$ and $y \in H^{2n-2m}(M, M - Y/G)$ are the Poincaré duals of ξ and η respectively then without loss of generality $j_X^*(x) = c^{n-k} \cdot 1_M$ and $j_Y^*(y) = c^{n-m} \cdot 1_M$. This implies

$$j_{X \cap Y}^*(x \smile y) = c^{2n-k-m} \cdot 1_M.$$

If $r := k + m - n \geq 1$ then $c^{2n-k-m} \cdot 1_M \neq 0$, hence by Poincaré duality

$$i_* \neq 0: H_{2r-1}\big((X \cap Y)/G\big) \longrightarrow H_{2r-1}(M).$$

The universal coefficient theorem now yields

$$c^{r-1}w \cdot 1_{X \cap Y} \neq 0 \in H_G^{2r-1}(X \cap Y) \cong H^{2r-1}\big((X \cap Y)/G\big).$$

Therefore $\ell_1(X \cap Y) \geq r + 1$ and $\ell(X \cap Y) \geq 2r$ as claimed.

  ad 2: With k, m and $r = k + m - n$ as above we shall show that

$$\ell(X \cap Y) \geq \ell(X) + \ell(Y) - \ell(SV) = 2k + 2m - 2n - 1 = 2r - 1.$$

This follows as in case 1 if we choose $\xi \in H_{2k-1}(X/G)$ and $\eta \in H_{2m-2}(Y/G)$. One obtains $j_{X \cap Y}^*(x \smile y) = c^{2n-k-m}w \cdot 1_M$ which implies $c^{r-1} \cdot 1_{X \cap Y} \neq 0 \in H_G^{2r-2}(X \cap Y)$ provided $r \geq 1$. Thus $\ell_0(X \cap Y) \geq r$ and therefore $\ell(X \cap Y) \geq 2r - 1$.
  □

Chapter 5

The length of representation spheres

5.1 Introduction

In this chapter we compute the $(\mathcal{A}, h^*, I)$-length ℓ of fixed point free representation spheres SV of G in certain special cases. For $\mathcal{A}$ we mainly use the set of non-trivial homogeneous G-spaces G/H, $H \subsetneq G$ a closed subgroup, because this is the most important one for applications. The cohomology theory h^* depends on the group and the ideal $I \subset h^*(\text{pt})$ will be chosen as to make the computation as easy as possible. In most cases $I = h^*(\text{pt})$.

The simplest groups from our point of view are tori and p-tori. We deal with them in §5.2. Here we can work with Borel cohomology: $h^* = H^*_G$. We get precisely the results we expect, that is $\ell(SV) = \dim V$ if G is a 2-torus and $\ell(SV) = \frac{1}{2} \dim V$ if G is a $(p\text{-})$torus $(p > 2)$. Using these computations we can compute $\mathcal{A}\text{-cat}(SV)$ and $\mathcal{A}\text{-genus}(SV)$ and prove those results which we have already mentioned in Chapter 2.

In §5.3 we consider cyclic p-groups: $G = \mathbb{Z}/(p^k)$. It turns out that Borel cohomology does not yield any interesting results if $k > 1$ whatever coefficients one chooses. Instead we work with equivariant K-theory K_G. Using these computations we get lower estimates for $\mathcal{A}\text{-cat}(SV)$ and $\mathcal{A}\text{-genus}(SV)$. The estimates are optimal in a certain sense. This indicates that equivariant K-theory is a good choice for cyclic p-groups.

The cohomology theory which seems to contain the most information is equivariant stable cohomotopy theory. The disadvantage of this theory is that it is difficult to compute. In §5.4 we use a closely related cohomology theory, namely stable cohomotopy applied to the Borel construction. This will simply be denoted by h^*. Here we are at least able to compute $h^*(SV)$ if $\dim V = \infty$, that is if SV is contractible. This suffices to prove Theorem 3.1a) which states that $\mathcal{A}\text{-cat}(SV) = \mathcal{A}\text{-genus}(SV) = \infty$ if G is p-toral. After a simple reduction to the case where G is a finite p-group we prove that $\ell(SV) = \infty$ using h^*. In fact, the contractibility of SV implies that $\ell(SV) = \ell(\text{pt})$ and one can prove that $\ell(\text{pt}) = \infty$. In §5.5 we study the question under what conditions $\ell(\text{pt})$ is infinite. We also address the problem of finding upper bounds for $\ell(X)$ if $X^G = \emptyset$ in terms of $\dim(X/G)$.

Sections 5.3 and 5.4 are the only places where we work with some explicit cohomology theories different from Borel cohomology. It is impossible to introduce these theories here in any detail and to prove the results which we quote. Therefore we only give references to the literature. A reader who is not familiar with equivariant K-theory or with Borel stable cohomotopy theory has to accept these results on faith. Still we believe it is instructive to see what kind of computations are necessary to compute the length of representation spheres for example in §5.3. We do not refer to these methods in later chapters.

5.2 Torus groups

In this section we consider the torus groups $(\mathbb{Z}/p)^k$ and $(S^1)^k$. V always denotes a finite-dimensional fixed point free representation of G. We want to compute the length of SV.

5.1 Theorem:
   Consider the torus group $G = (S^1)^k$, $k \geq 1$, and let ℓ denote $(\mathcal{A}, h^*, I)$-length with $\mathcal{A} = \{G/H : H \subsetneqq G$ a closed subgroup$\}$, $h^* = H_G^*(-; \mathbb{Q})$ Borel cohomology with rational coefficients, $I = R = h^*(\mathrm{pt})$. Then

$$(\mathcal{A}, h^*, I)\text{-length}(SV) = \frac{1}{2} \dim V.$$

5.2 Theorem:
   Consider the group $G = (\mathbb{Z}/2)^k$, $k \geq 1$, and let ℓ denote $(\mathcal{A}, h^*, I)$-length with $\mathcal{A} = \{G/H : H \subsetneqq G$ a closed subgroup$\}$, $h^* = H_G^*(-; \mathbb{F}_2)$ Borel cohomology with coefficients in the field $\mathbb{F}_2$ of two elements, $I = R = h^*(\mathrm{pt})$. Then

$$(\mathcal{A}, h^*, I)\text{-length}(SV) = \dim V.$$

5.3 Theorem:
   Consider the group $G = (\mathbb{Z}/p)^k$, $k \geq 1$, and let ℓ_0 denote $(\mathcal{A}, h^*, I_0)$-length with $\mathcal{A} = \{G/H : H \subsetneqq G$ a closed subgroup$\}$, $h^* = H_G^*(-; \mathbb{F}_p)$ Borel cohomology with coefficients in $\mathbb{F}_p$, $I_0 = (c_1, \ldots, c_k) \subset \mathbb{F}_p[c_1, \ldots, c_k] \otimes E[w_1, \ldots, w_k] = R = h^*(\mathrm{pt})$. Then

$$(\mathcal{A}, h^*, I_0)\text{-length}(SV) = \frac{1}{2} \dim V.$$

5.4 Remark:
   a) If one chooses other ideals of R in the case $G = (\mathbb{Z}/p)^k$ then the length $\ell(SV)$ becomes slightly more complicated to describe. We shall only mention the case $k = 1$ and $I_1 = R$ as in Example 4.5. Here one obtains

$$\ell_1(SV) = 1 + \ell_0(SV) = 1 + \frac{1}{2} \dim V.$$

   b) If $G = \mathbb{Z}/p$ and ℓ is defined as in Remark 4.14 then $\ell(SV) = \dim V$. Moreover, for this ℓ it follows that $\ell(J_r G) = r$ where $J_r G$ is the join of r copies of G. To see this, one first constructs G-maps $G * G \to S^1$ and $S^1 \to G * G$ where G acts on S^1 via complex multiplication. This implies $\ell_0(J_{2r} G) = \ell_0(S^{2r-1}) = r$ and $\ell_1(J_{2r} G) = \ell_1(S^{2r-1}) = r + 1$, hence $\ell(J_{2r} G) = 2r$. The subadditivity property yields $\ell(J_{2r+1} G) \leq 2r + 1$ because $\ell(G) = 1$. Applying subadditivity once more we obtain equality:

$$4r + 2 = \ell(J_{4r+2} G) = \ell(J_{2r+1} G * J_{2r+1} G) \leq 2\ell(J_{2r+1} G).$$

Observe that the additional assumptions which we need to apply subadditivity to ℓ are satisfied in both cases; cf. Remark 4.14.

The theorems are all proved in the same way. One computes $h^*(SV)$ with the help of a Gysin sequence and can then check the definition of the length explicitly. Recall that the *Gysin sequence* of an oriented vector bundle $E \to B$ with fibre $\mathbb{R}^n$ has the following form:

$$\cdots \longrightarrow H^{i-n}(B) \xrightarrow{\ e\ } H^i(B) \longrightarrow H^i(SE) \longrightarrow H^{i+1-n}(B) \xrightarrow{\ e\ } \cdots$$

Here SE is the unit sphere bundle of E (for some Riemannian metric on E) and $\xrightarrow{\ e\ }$ denotes multiplication with the Euler class $e \in H^n(B)$ of the vector bundle. The coefficients of H^* are arbitrary. If $E \to B$ is not orientable then the Euler class and the Gysin sequence also exist provided one takes cohomology with coefficients in $\mathbb{F}_2$. The Euler class of the Whitney sum $E \oplus E'$ is the product $e \cdot e'$ of the Euler classes of E and E'. Standard references for this are [MilS] or [Sp].

The following simple observation is helpful for computing the $(\mathcal{A}, h^*, I)$-length.

5.5 Observation:
If $\mathcal{A}' \subset \mathcal{A}$ is such that for each $A \in \mathcal{A}$ there exists $A' \in \mathcal{A}'$ and a G-map $A \to A'$ then

$$(\mathcal{A}, h^*, I)\text{-length} = (\mathcal{A}', h^*, I)\text{-length}.$$

Here "$\leq$" is a consequence of $\mathcal{A}' \subset \mathcal{A}$ and "$\geq$" follows immediately from the definition because

$$\ker\big(R \to h^*(\mathcal{A}')\big) \subset \ker\big(R \to h^*(\mathcal{A})\big).$$

Proof of Theorem 5.1:
First of all, because of 5.5 we may replace $\mathcal{A}$ by $\mathcal{A}' = \{G/H \in \mathcal{A} : \operatorname{rank}(H) = k - 1\}$. The Künneth theorem yields $R \cong \mathbb{Q}[c_1, \ldots, c_k]$ where $c_i \in h^2(\mathrm{pt})$ comes from the projection $G \to S^1$ on the i-th factor. Given $G/H \in \mathcal{A}'$ there exists a homomorphism $\chi_H : G \to S^1$ such that $\ker \chi_H = H$; χ_H is not unique, of course. Let $V_H \cong \mathbb{R}^2$ be the irreducible representation of G corresponding to the character χ_H. Then the unit sphere SV_H and the homogeneous space G/H are homeomorphic G-spaces. The Gysin sequence of the bundle $EG \underset{G}{\times} V_H \to BG$ with Euler class denoted by $e(V_H) \in H^2(BG) \cong h^2(\mathrm{pt}) \subset R$ looks as follows:

$$\cdots \longrightarrow h^{i-2}(\mathrm{pt}) \xrightarrow{e(V_H)} h^i(\mathrm{pt}) \longrightarrow h^i(SV_H) \longrightarrow h^{i-1}(\mathrm{pt}) \xrightarrow{e(V_H)} \cdots$$

Since $R = h^*(\mathrm{pt})$ has no zero divisors, multiplication with $e(V_H)$ is injective. Therefore $h^*(SV_H) \cong R/\big(e(V_H)\big)$ and $\ker\big(R \to h^*(SV_H)\big) = \big(e(V_H)\big)$. Next we study the map $p^* : h^*(\mathrm{pt}) \to h^*(SV)$ induced by the projection. As above we obtain $h^*(SV) \cong R/\big(e(V)\big)$ which implies that p^* is injective in dimensions less than $n = \dim V$. Consequently, $\ell(V) \geq m := n/2$. Now V splits as the sum $V_1 \oplus \ldots \oplus V_m$ of m irreducible representations V_i and, hence, the Euler class $e(V)$ is the product $e(V_1) \cdot \ldots \cdot e(V_m)$ of the Euler classes $e(V_i)$. We have seen above that the SV_i, $i = 1, \ldots, m$, can be considered as elements of $\mathcal{A}'$ and that $\ker\big(R \to h^*(SV_i)\big) = \big(e(V_i)\big)$. This yields $\ell(SV) \leq m$ because the product $e(V_1) \cdot \ldots \cdot e(V_m) = e(V)$ maps to zero in $h^*(SV)$.□

Proof of Theorem 5.2:

The proof is completely analogous to that of 5.1. Here

$$R = h^*(\text{pt}) \cong \mathbb{F}_2[w_1, \ldots, w_k]$$

with $w_i \in h^1(\text{pt})$. The irreducible (real) representations are all one-dimensional. Thus the factor $1/2$ does not appear. □

Proof of Theorem 5.3:

Again the proof proceeds as the one of 5.1 although the coefficient ring is not a polynomial ring. But due to our choice of the ideal I_0 we get for $G/H \in \mathcal{A}'$:

$$I_0 \cap \ker\left(R \to h^*(G/H)\right) \subset I_0 = (c_1, \ldots, c_k).$$

So for the computation of the length we only have to consider products of the polynomial generators c_i. □

The proof of Theorem 5.3 also shows what one has to do when I_0 is replaced by other ideals of R. For instance, if $G = \mathbb{Z}/p$ and $I = R \cong \mathbb{F}_p[c] \otimes E[w]$ then $\ker\left(R \to h^*(\mathbb{Z}/p)\right) \cong (c, w)$. Furthermore, the Euler class of a non-trivial irreducible representation $V_j \cong \mathbb{C}$ is given by jc if the character of V_j is $G \ni \zeta \mapsto \zeta^j \in S^1$. Therefore $h^*(SV) \cong R/(c^m)$ with $m = \frac{1}{2} \dim V$. Consequently, $c^{m-1}w \neq 0 \in h^*(SV)$ which implies $\ell_1(SV) = 1 + \frac{1}{2} \dim V$.

5.6 Remark:

As a consequence of these computations we obtain the following generalization of the Borsuk-Ulam theorem.

If V and W are fixed point free representations of $G = (\mathbb{Z}/p)^k$ or $G = (S^1)^k$ then the existence of a G-map $SV \to SW$ implies $\dim W \geq \dim V$.

We also obtain for $G = (\mathbb{Z}/p)^k$ and $\mathcal{A} = \{G/H: H \subsetneq G \text{ a subgroup}\}$:

$$\mathcal{A}\text{-cat}(SV) = \mathcal{A}\text{-genus}(SV) = \dim V;$$

and for $G = (S^1)^k$ and $\mathcal{A} = \{G/H: H \subsetneq G \text{ a closed subgroup}\}$:

$$\mathcal{A}\text{-cat}(SV) = \mathcal{A}\text{-genus}(SV) = \frac{1}{2} \dim V.$$

Of course, $V^G = 0$ is understood. If p is an odd prime then one needs a little trick to get rid of the factor $1/2$ in the computation of $\mathcal{A}$-cat and $\mathcal{A}$-genus. Namely, if $\mathcal{A}$-genus$(SV) = k$ then there exists a G-map $SV \to G/H_1 * \ldots * G/H_k$. The join of this map with itself yields a G-map

$$S(V \oplus V) \longrightarrow (G/H_1 * G/H_1) * \ldots * (G/H_k * G/H_k).$$

We may assume that $G/H_i \cong \mathbb{Z}/p$. Then it is not difficult to construct a G-map $G/H_i * G/H_i \to SV_i$ where $V_i \cong \mathbb{C}$ is an irreducible fixed point free representation of G corresponding to a character $G \to G/H_i \hookrightarrow S^1$. Thus we obtain a G-map

$S(V \oplus V) \to SW$ with $\dim W = 2k$. The above Borsuk-Ulam theorem implies $2k = \dim W \geq \dim(V \oplus V)$. This yields

$$\mathcal{A}\text{-cat}(SV) \geq \mathcal{A}\text{-genus}(SV) \geq \dim V.$$

The other inequality can easily be seen geometrically.

It is possible to prove these results without the help of cohomology theory. An elementary proof using only the Brouwer degree and Sard's lemma can be found in [Ba4].

5.3 Cyclic p-groups

We consider the group $G = \mathbb{Z}/(p^k)$. First we show that Borel cohomology is not very helpful in this case.

5.7 Example:

We assume $k > 1$ so that G has subgroups different from e or G. To compute $\ell = (\mathcal{A}, H_G^*, I)$-length with $\mathcal{A} \subset \{G/H \colon H \subsetneqq G$ a closed subgroup$\}$ it is no loss of generality to assume that $\mathcal{A}$ consists of just one element G/H because the subgroups of G are totally ordered (see 5.5). Moreover we suppose that $H = \mathbb{Z}/(p^m)$ with $1 \leq m < k$. We do not consider the case $m = 0$ because $\mathcal{A} = \{G\}$ is only useful for free G-spaces, a situation which can always be reduced to $\mathbb{Z}/p$. If the action of G on X is not free then $\{G\}$-cat$(X) = \{G\}$-genus$(X) = \infty$. The length will also depend on the choice of the coefficients. We use either integer coefficients or $\mathbb{F}_p$ but other choices will not improve the result. Finally we set $I = R$.

For any natural number n the inclusion $\mathbb{Z}/n \hookrightarrow S^1$ induces a ring epimorphism

$$\mathbb{Z}/[c] \cong H^*(BS^1; \mathbb{Z}) \longrightarrow H^*(B\mathbb{Z}/n; \mathbb{Z}) \cong \begin{cases} \mathbb{Z} & \text{if } * = 0; \\ \mathbb{Z}/n & \text{if } * > 0 \text{ is even}; \\ 0 & \text{if } * > 0 \text{ is odd}. \end{cases}$$

In positive degrees this is just the reduction modulo n. Of course, in degree 0 it is an isomorphism $H^0(BS^1; \mathbb{Z}) \cong H^0(B\mathbb{Z}/n; \mathbb{Z})$. Consequently, the map $\alpha \colon G/H \to$ pt induces an epimorphism $\alpha^* \colon R = H_G^*(\text{pt}; \mathbb{Z}) \longrightarrow H_G^*(G/H; \mathbb{Z})$ which is the reduction modulo p^m in positive degrees. Therefore the kernel of α^* is generated by $p^m \cdot c$, that is $\ker(\alpha^*) = p^m \cdot c \cdot R$. Setting $\langle k/m \rangle := \min\{n \in \mathbb{Z} \colon n \geq k/m\}$ we see that $(p^m \cdot c)^{\langle k/m \rangle}$ is divisible by p^k hence it is zero in R. This shows that $\ell(\text{pt}) \leq \langle k/m \rangle$, and in fact, equality holds. Therefore $\ell(X) \leq \langle k/m \rangle$ for every G-space X according to Corollary 4.8. If we have to deal with fixed point free G-spaces X where $H = \mathbb{Z}/(p^{k-1})$ occurs as isotropy group then we have to use $\mathcal{A} = \{G/H\}$. In this case we only obtain

$$\mathcal{A}\text{-cat}(X) \geq \mathcal{A}\text{-genus}(X) \geq \ell(X) \in \{1, 2\}.$$

The estimate is even worse if we use $\mathbb{F}_p$-coefficients. The following results can be found in [FiP], §VI.2. We obtain

$$H^*(B\mathbb{Z}/(p^m); \mathbb{F}_p) \cong \mathbb{F}_p[c] \otimes E[w]$$

independently of $m \geq 1$ (except if $p = 2$ and $m = 1$ when $H^*(B\mathbb{Z}/2; \mathbb{F}_2) \cong \mathbb{F}_2[w]$). As usual, the degree of c is 2 and the degree of w is 1. The map $\alpha\colon G/H \to \text{pt}$ induces a homomorphism $\alpha^*\colon R = H_G^*(\text{pt}; \mathbb{F}_p) \longrightarrow H_G^*(G/H; \mathbb{F}_p)$ which is the identity on $\mathbb{F}_p[c]$ and maps w to 0 if $H \subsetneqq G$. Thus the kernel of α^* is generated by w. But since $w^2 = 0$ we obtain $\ell(\text{pt}) = 2$ for every $\mathcal{A} \subset \{G/H\colon H \subsetneqq G \text{ a closed subgroup}\}$.

One obtains better results using *equivariant K-theory K_G* instead. $K_G(X)$ is the Grothendieck group of isomorphism classes of complex G-vector bundles over X. For example, the coefficient ring $R = K_G(\text{pt}) \cong R(G)$ is the complex representation ring of G. All facts which we need to know about K_G can be found in [At] and [Sel]. The following theorem and its consequences are illustrating even without prior knowledge of K_G. Of course, then one has to accept the results quoted in the proof on faith. We fix some notation. Given two powers $1 \leq m \leq n < p^k$ of p we set

$$\mathcal{A}_{m,n} = \{G/H\colon H \subset G, \ m \leq |H| \leq n\}$$

and $\ell_n = (\mathcal{A}_{m,n}, K_G, R)$-length. This is independent of m because of 5.5. The notation does not quite coincide with the one of chapter 4. One should replace K_G by the associated cohomology theory K_G^* and use the ideal $I := K_G(\text{pt}) \subset K_G^*(\text{pt})$. Then $\ell_n = (\mathcal{A}_{m,n}, K_G^*, I)$-length would be the correct notation. Recall that we write $\mathcal{A}_X$ for a set consisting of all G-orbits of X (up to homeomorphism).

5.8 Theorem:

Let V be a complex representation of $G = \mathbb{Z}/(p^k)$ and set $d := \frac{1}{2}\dim V$. Fix two powers m, n of p as above. Then

$$\ell_n(SV) \geq \begin{cases} 1 + \left\langle \frac{(d-1)m}{n} \right\rangle & \text{if } \mathcal{A}_{SV} \subset \mathcal{A}_{m,n}, \\ \infty & \text{if } \mathcal{A}_{SV} \not\subset \mathcal{A}_{1,n}. \end{cases}$$

Here $\langle x \rangle$ is the least integer greater than or equal to x. Moreover, if $\mathcal{A}_{SV} \subset \mathcal{A}_{n,n}$ then

$$\ell_n(SV) = d = \frac{1}{2}\dim V.$$

Postponing the proof of 5.8 to the end of this section we first collect a number of consequences and related results. Theorem 5.8 (and its proof) show once more how important the knowledge of the orbit types is for the computation of the length. Of course, if we know precisely all orbits appearing in SV then we may define $\mathcal{A}$ accordingly. The proof will show that the estimate of $\ell_n(SV)$ can be improved in this case, but the computations are intricate. In a concrete application it may be difficult to determine the orbit structure, for instance if V appears as the kernel of the Hessian of a potential operator. Observe that in the case $G = \mathbb{Z}/p$ (that is $k = m = n = 1$) equivariant K-theory yields the same result as Borel cohomology (see Theorem 5.3).

5.9 Corollary:

Let V and W be fixed point free representations of G.
a) *If there exists a G-map $SV \to SW$ then $\dim W \geq (\dim V)/p^{k-1}$.*
b) *If $\mathcal{A}_{SV} \subset \mathcal{A}_{m,n}$ for some powers $1 \leq m \leq n < p^k$ of p then*

$$\mathcal{A}_{m,n}\text{-cat}(SV) \geq \mathcal{A}_{m,n}\text{-genus}(SV) \geq \frac{m}{n} \dim V.$$

Proof:

a) Even if $p = 2$ we may assume that V and W are complex representations. If not we work instead with $V \otimes \mathbb{C}$ and $W \otimes \mathbb{C}$ and pass to the join of the G-map $SV \to SW$. This is a G-map

$$S(V \otimes \mathbb{C}) \cong SV * SV \to SW * SW \cong S(W \otimes \mathbb{C}).$$

Moreover, it is no loss of generality to assume that all orbits of SW have isotropy $\mathbb{Z}/p^{k-1}$ because there always exists a G-map $SW \to SW'$ where $\zeta \in G \subset S^1$ acts on $W' \cong \mathbb{C}^d$, $\dim W' = \dim W$, via multiplication with $\zeta^{p^{k-1}}$. Setting $n = p^{k-1}$ Theorem 5.8 and the monotonicity of ℓ yield

$$\dim W = 2 \cdot \ell_n(SW) \geq 2 \cdot \ell_n(SV) \geq (\dim V)/p^{k-1}.$$

b) It is clear that

$$\mathcal{A}_{m,n}\text{-cat}(SV) \geq \mathcal{A}_{m,n}\text{-genus}(SV) = \mathcal{A}_{n,n}\text{-genus}(SV) =: \gamma.$$

Let $f: SV \to A_1 * \ldots * A_\gamma$ with $A_i = G/H \in \mathcal{A}_{n,n}$ be a G-map according to the definition of the genus. It is easy to construct a G-map $A_i * A_i \to SW$ where $\zeta \in G$ acts on $W \cong \mathbb{C}$ via scalar multiplication with ζ^n. Then $f * f$ induces a G-map

$$S(V \oplus V) \cong SV * SV \longrightarrow A_1 * A_1 * \ldots * A_\gamma * A_\gamma \longrightarrow SW^\gamma.$$

Theorem 5.8 implies

$$\gamma = \ell_n(SW^\gamma) \geq \ell_n\big(S(V \oplus V)\big) \geq \frac{m}{n} \dim V.$$

$\square$

5.10 Remark:

Of course, one can now ask whether this "weak" estimate in Corollary 5.9 can be improved. This is certainly true for part a) if $\mathcal{A}_{SW} \subset \mathcal{A}_{m,n}$ with n small. Observe that the proof of Corollary 5.9a) yields a slightly better result if V and W are complex representations. Then

$$\dim W \geq 2 \left\langle \left(\frac{1}{2} \dim V - 1 \right) \Big/ p^{k-1} \right\rangle + 2 \geq \langle (\dim V)/p^{k-1} \rangle$$

if there exists a G-map $SV \to SW$. However, Theorems 3.23 and 3.24 show that the factors $1/p^{k-1}$ and m/n are optimal and not just caused by equivariant K-theory. Consider for instance the simplest non-trivial case $G = \mathbb{Z}/4$. Theorem 3.23a) tells us that there exist G-maps $SV \to SW$ with $\dim W = 1 + (\dim V)/2$ if $\dim V \equiv 0, 4$

(mod 8). Here G acts freely on SV and via the antipodal map on SW. According to Corollary 5.9 $\dim W \geq (\dim V)/2$. So the factor $1/2$ is optimal. Similarly, using the monotonicity of the genus we see that for V and W as above

$$\frac{1}{2}\dim V \leq \mathcal{A}_{1,2}\text{-genus}(SV) \leq \mathcal{A}_{1,2}\text{-genus}(SW) = \dim W = 1 + \frac{1}{2}\dim V.$$

Let α_G be the largest real number satisfying $\mathcal{A}_{SV}\text{-genus}(SV) \geq \alpha_G \cdot \dim V$. It is not difficult to see that for finite abelian groups α_G coincides with the number a_G defined in Theorem 3.24: a_G is the largest real number such that the existence of a G-map between fixed point free representation spheres $SV \to SW$ implies $\dim W \geq a_G \cdot \dim V$. We showed in 3.24b) that $a_G = 1/p^{k-1}$ for $G = \mathbb{Z}/(p^k)$. Thus $\alpha_G = a_G = 1/p^{k-1}$ if $G = \mathbb{Z}/p^k$ and the factor m/n in Corollary 5.9b) is optimal as far as the genus is concerned.

For the category we do not know whether 5.9b) is optimal. Let $\beta_G \geq 0$ be the maximal real number satisfying $\mathcal{A}_{1,n}\text{-cat}(SV) \geq \beta_G \cdot \dim V$ for all fixed point free representations V of G. According to 3.24b) there exist G-maps $SV \to SW$ with $\dim W = (\dim V)/p^{k-1} + o(\dim V)$. Using the weak monotonicity of the category in the sense of Proposition 2.18 we can only conclude ($n = p^{k-1}$):

$$\frac{1}{n}\dim(V \oplus W) \leq \mathcal{A}_{1,n}\text{-cat}\big(S(V \oplus W)\big) \leq 2 \cdot \mathcal{A}_{1,n}\text{-cat}(SW)$$

$$\leq 2 \cdot \dim W = \frac{2}{1+n}\dim(V \oplus W) + o(\dim V).$$

This implies $1/p^{k-1} \leq \beta_G \leq 2/(1 + p^{k-1})$ for $G = \mathbb{Z}/p^k$.

Proof of Theorem 5.8:

Set $H := \mathbb{Z}/n \subset G$. The coefficient ring $R = K_G(\mathrm{pt}) \cong R(G)$ is isomorphic to $\mathbb{Z}[x]/(1 - x^{p^k})$. Similarly, $K_G(G/H) \cong K_H(\mathrm{pt}) \cong R(H)$ is isomorphic to $\mathbb{Z}[x]/(1 - x^n)$. So the kernel of the homomorphism $R \to K_G(G/H)$ is the ideal $(1 - x^n) \subset R$. The Gysin sequence of a complex G-module W yields the exact sequence

$$R \xrightarrow{e_W} R \twoheadrightarrow K_G(SW) \to 0.$$

The first map is multiplication with the Euler class $e_W \in R$ of the complex G-vector bundle $W \to \mathrm{pt}$. Consequently, $K_G(SW) \cong R/(e_W)$. Since the Euler class is multiplicative we only have to compute the Euler classes of the irreducible representations. For $i = 0, 1, \ldots, p^k - 1$ let $\zeta \in G$ act on $V_i \cong \mathbb{C}$ via scalar multiplication with ζ^i. The representation V_i corresponds to the monomial $x^i \in \mathbb{Z}[x]/(1 - x^{p^k}) \cong R$. The Euler class e_{V_i} is given by $1 - x^i$. This corresponds to the element $V_0 - V_i$ in $R(G)$. In general, the Euler class of W corresponds to the element $\sum_{j=0}^{\dim W}(-1)^j \Lambda^j W$ where $\Lambda^j W$ is the j-th exterior power of W. Next we decompose V into its irreducible summands:

$$V \cong V_0^{n_0} \oplus \ldots \oplus V_r^{n_r}, \quad \text{where } r = p^k - 1.$$

The above discussion of the Euler classes shows that

$$e_V = (1 - x^0)^{n_0} \cdot \ldots \cdot (1 - x^r)^{n_r}$$

and $K_G(SV) \cong \mathbb{Z}[x]/(1 - x^{p^k}, e_V)$. Now we can compute $\ell_n(SV)$: $\ell_n(SV)$ is the smallest integer $l \geq 0$ such that $(1 - x^n)^l$ is contained in the ideal $(1 - x^{p^k}, e_V)$ of $\mathbb{Z}[x]$. We distinguish two cases. First suppose there exists an orbit of SV which is not contained in $\mathcal{A}_{1,n}$. This means that $n_i > 0$ for some multiple i of np. So e_V is divisible by $1 - x^{np}$, hence, $(1 - x^n)^l$ must be divisible by $1 - x^{np}$. This is a contradiction. Therefore such an l cannot exist, that is $\ell_n(SV) = \infty$. It remains to consider the case where $\mathcal{A}_{SV} \subset \mathcal{A}_{m,n}$. This is equivalent to $V \cong \bigoplus_{j=0}^r V_j^{n_j}$, where $n_j = 0$ if j is not divisible by m or if j is divisible by pn. This implies that e_V is divisible by $(1 - x^m)^d$ with $d = (\dim V)/2 = \sum_{j=0}^r n_i$. Therefore the ideal $(1 - x^{p^k}, e_V)$ is contained in the ideal $(1 - x^{pn}, (1 - x^m)^d)$, and $(1 - x^n)^l$ is an element of it; $l = \ell_n(SV)$. Setting $y = x^m$ we obtain that $(1 - y^{n/m})^l$ is an element of the ideal $(1 - y^{pn/m}, (1 - y)^d)$. Now Proposition 3.1 of [Ba2] tells us that this can only happen if

$$(\frac{n}{m} - 1) \cdot (l - 1) \geq d - l$$

or $l \geq 1 + (d - 1)m/n$ as claimed. $\square$

5.4 Proof of Theorem 3.1a)

Let G be a p-toral group and X be a contractible fixed point free G-ANR. We want to show that

$$\mathcal{A}_X\text{-cat}(X) = \mathcal{A}_X\text{-genus}(X) = \infty.$$

In fact we prove a slightly more general statement. We claim that $\mathcal{A}$-genus$(X) = \infty$ where $\mathcal{A}$ is any set of compact fixed point free G-spaces. For instance we allow $\mathcal{A} = \widehat{\mathcal{A}}_X$. Suppose to the contrary that $k := \mathcal{A}$-genus$(X) < \infty$. By definition there exist $A_1, \ldots, A_k \in \mathcal{A}$ and a G-map

$$X \longrightarrow A_1 * \ldots * A_k =: A.$$

We claim that there exists a finite p-group P of G which acts on A without fixed points: $A^P = \emptyset$. To see this observe that G can be approximated by finite p-groups. More precisely, for any natural number s consider the set

$$P_s := \{g \in G : g^{p^s} = e\}.$$

If p^s is a multiple of the order of G/G_0 then P_s is a subgroup of G according to [Henn], Lemma 1.4. P_s is obviously a finite p-group. It is an extension of $G_0 \cap P_s$ by G/G_0. Moreover, it is clear that $A^{P_s} = \emptyset$ if s is big enough because A is a compact fixed point free G-space. Setting $P = P_s$ for such an s we consider X and A as P-spaces and $X \to A$ as P-equivariant. In order to prove that such a P-map cannot exist we apply an appropriate version of the length. The P-cohomology theory which we use is *Borel stable cohomotopy*:

$$h^*(X, Y) := \pi_s^*(EP \underset{P}{\times} X, EP \underset{P}{\times} Y) \quad \text{where } \pi_s^* \text{ is stable cohomotopy theory.}$$

See [Wh], Chapter XII, for the definition of π_s^*. Setting $\mathcal{B} = \{P/H : H \subsetneq P\}$ and $I = h^*(\text{pt})$ Corollary 4.9 implies $\ell(A) < \infty$ where $\ell = (\mathcal{B}, h^*, I)$-length. We shall prove below that $\ell(X) = \infty$. Thus there cannot exist a P-map $X \to A$ because of the monotonicity property of the length. $\ell(X) = \infty$ is a consequence of the following two lemmas.

5.11 Lemma:
The projection $X \to$ pt induces a homotopy equivalence $EP \underset{P}{\times} X \to BP$, hence an isomorphism $h^*(\text{pt}) \xrightarrow{\cong} h^*(X)$.

5.12 Lemma:
There exists $\omega \in h^0(\text{pt})$ such that $\omega \in \ker\left(h^0(\text{pt}) \to h^0(P/H)\right)$ for all $P/H \in \mathcal{B}$. Moreover, ω is not nilpotent: $\omega^n \neq 0$ for all $n \in \mathbb{N}$.

Lemma 5.11 implies $\ell(X) = \ell(\text{pt})$ and Lemma 5.12 implies $\ell(\text{pt}) = \infty$. Observe that we only need h^0 and not the full cohomology theory h^*.

Proof of Lemma 5.11:
Since X is contractible the homotopy sequence of the fibration $\pi: EP \underset{P}{\times} X \to BP$ with fibre X shows that π induces isomorphisms for all homotopy groups. Since $EP \underset{P}{\times} X$ and BP have the homotopy types of CW-complexes Whitehead's theorem tells us that π is a homotopy equivalence. $\quad\square$

Proof of Lemma 5.12:
This is more complicated. Due to Carlsson's solution of the Segal conjecture (see [Ca]), $h^0(\text{pt})$ is isomorphic to the completion $\widehat{\Omega}(P)$ of the Burnside ring $\Omega(P)$ by the augmentation ideal $I(P)$. This can be phrased in terms which are already familiar to us. Let $V_0, \ldots, V_n$ be a complete set of complex irreducible representations of P and set $V := \bigoplus_{i=0}^n V_i \oplus V_i$. Then $\Omega(P) \cong [SV, SV]^P$ where the right hand side denotes the set of P-homotopy classes of P-maps $SV \to SV$; all maps are required to preserve the base point in SV^P (see [Se2] or [Di]). Under this isomorphism the ideal $I(P)$ corresponds to the kernel of the homomorphism $[SV, SV]^P \to \mathbb{Z}$, $[f] \mapsto \deg(f)$. If H is a subgroup of P then we can consider V as a representation of H and $\Omega(H) \cong [SV, SV]^H$. The completion homomorphism

$$[SV, SV]^H \cong \Omega(H) \longrightarrow \widehat{\Omega}(H) \cong \pi_s^0(BH) \cong h^0(P/H)$$

is injective for every $H \subset P$ because P is a p-group (see [Lai], Corollary 1.11). Now we use the fact that the ring homomorphism

$$[SV, SV]^H \longrightarrow \prod_{K \subset H} \mathbb{Z}, \quad [f] \mapsto (\deg(f^K): K \subset H),$$

is injective for every H (see [Di], Theorem II(4.11)). Consequently, $[SV, SV]^P$ has no zero divisors, in particular no nilpotent elements. Because of Proposition 3.10 there exists a P-map $f: SV \to SV$ such that

$$\deg(f^H) = \begin{cases} 0 & \text{if } H \subsetneq P, \\ |P| & \text{if } H = P. \end{cases}$$

This implies that $\alpha := [f] \in \ker([SV, SV]^P \to [SV, SV]^H)$ for any $H \subsetneq P$. This element $\alpha \in \Omega(P)$ exists for every finite group. P need not be a p-group. Let $\omega \in h^0(\text{pt})$ be the image of $\alpha \in \Omega(P)$ under the completion homomorphism. Since this homomorphism is injective ω is not nilpotent. Here we need that P is a p-group. For non-p-groups the completion homomorphism is not injective (see [Lai]). Moreover, $\omega \in \ker \left(h^0(\text{pt}) \to h^0(P/H) \right)$ because the following diagram is commutative:

$$
\begin{array}{ccc}
[SV, SV]^P & \longrightarrow & [SV, SV]^H \\
\downarrow & & \downarrow \\
h^0(\text{pt}) & \longrightarrow & h^0(P/H)
\end{array}
$$

The vertical maps are the completion homomorphisms.   □

Theorem 3.1a) has essentially been proved in [BCP]. Our proof is a simplified version of the one in [BCP]. Unfortunately, this approach does not yield any lower bound for the category, genus or length of finite-dimensional representation spheres of a p-group. Probably one has to use Borel stable cohomotopy or equivariant stable cohomotopy theory. But the computation of the Euler classes seems to be difficult.

5.5 Related results

In the last section 5.4 we have used in an essential way that $\ell(\text{pt}) = \infty$ if G is a p-group and ℓ is defined with Borel stable cohomotopy theory. We also saw in Example 5.7 that $\ell(\text{pt})$ can be finite even for cyclic p-groups if we work with the wrong cohomology theory. We want to pursue this problem a little further without introducing too many new cohomology theories.

5.13 Proposition:
Let G be a compact Lie group which is not p-toral and let $\mathcal{A}$ be the set of all nontrivial homogeneous G-spaces G/H, $H \subsetneq G$ a closed subgroup. Then $\ell(\text{pt}) < \infty$ for any $\ell = (\mathcal{A}, h^*, I)$-length where h^* is a multiplicative equivariant cohomology theory of Borel type. By this we mean that $h^*(X) = k^*(EG \underset{G}{\times} X)$ for some multiplicative (generalized) cohomology theory k^*.

Proof:
Let X be a (non-equivariantly) contractible fixed point free G-ANR satisfying $\mathcal{A}_X$-cat$(X) < \infty$. Such a space exists because of Theorem 3.1b). Then the projection $X \to \text{pt}$ induces a homotopy equivalence $EG \underset{G}{\times} X \to BG$; compare Lemma 5.11. Thus we obtain (using Proposition 4.2)

$$
\ell(\text{pt}) = \ell(X) \le \mathcal{A}_X\text{-cat}(X) < \infty
$$

because $h^*(\text{pt}) \cong h^*(X)$ for a Borel type theory.   □

In Proposition 5.13 it is essential that h^* is defined via a Borel construction. To see this consider a finite group G and let ω^* denote equivariant stable cohomotopy theory (see [Se2]). In particular,

$$\omega^0(\text{pt}) \cong \Omega(G) \cong [SV, SV]^G$$

where $\Omega(G)$ is the Burnside ring of G and V contains two copies of every complex irreducible representation as in the proof of Lemma 5.12.

5.14 Proposition:
Let G be a finite group, let $\mathcal{A}$ be any set of compact fixed point free G-spaces and set $R = \omega^*(\text{pt})$. Then $\ell(\text{pt}) = \infty$ for $\ell = (\mathcal{A}, \omega^*, R)$-length.

Proof:
As mentioned in the proof of Lemma 5.12 there exists a non-zero element $\alpha \in \Omega(G) = \omega^0(\text{pt})$ which lies in the kernel of every homomorphism $\omega^0(\text{pt}) \longrightarrow \omega^0(G/H)$ provided $H \subsetneqq G$. Moreover, α is not nilpotent. Thus for every $A \in \mathcal{A}$ there exists an integer $k = k(A) \geq 1$ with $\alpha^k \in \ker\big(\omega^0(\text{pt}) \to \omega^0(A)\big)$ because

$$(\mathcal{B}, \omega^*, R)\text{-length}(A) < \infty \qquad \text{where } \mathcal{B} = \{G/H \colon H \subsetneqq G \text{ a subgroup}\};$$

see Corollary 4.9b). Now $\ell(\text{pt}) = \infty$, again because α is not nilpotent. $\square$

Proposition 5.14 is false if G is a torus because $\omega^0(\text{pt}) \longrightarrow \omega^0(G/H)$ is zero for every proper closed subgroup H of G. Equivariant stable cohomotopy is most suitable for finite groups, not so much for infinite compact Lie groups. These two propositions illustrate once more how important the choice of the cohomology theory is. On the other hand, one's choice is restricted by the possibility of doing calculations. One can only work with those theories h^* for which one can compute at least the ring structure of $h^0(\text{pt})$ or $h^*(\text{pt})$ and some Euler classes. This can already be very difficult.

We have seen in Chapter 4 that many properties of the category and the genus hold for the length, too. The proofs of these results in §4.4 needed only the general properties of multiplicative cohomology theories and were almost straightforward. There are two properties of the category and the genus which are useful for computations and which we have not generalized to the length. The first is an estimate of $\mathcal{A}$-cat(X) and $\mathcal{A}$-genus(X) from above using the dimension of X/G as in Propositions 2.13b) and 2.16b). The second is a relation between the category or genus of X as a G-space and the category or genus of X as an H-space for some subgroup H of G; see Propositions 2.14 and 2.17. It is not at all clear whether one can generalize these results to the length in a non-trivial way. Here we shall only prove one result in this direction which deals with a very special case. Another such instance will be studied in §8.4 where we are interested in the length of $O(3)$-spaces considered as $\mathbb{Z}/2$-spaces. The subgroup $\mathbb{Z}/2$ of $O(3)$ consists of $\{\pm 1\}$.

5.15 Proposition:
   Let T be the maximal torus of G and let $\ell := (\mathcal{A}, H_T^*, R)$-length be the length defined in Example 4.3 for T. Then for any closed subgroup H of G:

$$\ell(G/H) = \infty \qquad\qquad if \operatorname{rk}(H) = \operatorname{rk}(G) := \dim T;$$

and

$$\ell(G/H) \le |W| \cdot |G/HG_0| \quad if \operatorname{rk}(H) < \operatorname{rk}(G).$$

Here G_0 is the component of the unit element of G and $W = N_{G_0}T/T$ is the Weyl group.

5.16 Remark:
   a) We set $\mathcal{B} = \{G/H\colon H$ a closed subgroup of G and $\operatorname{rk}(H) < \operatorname{rk}(G)\}$ and write ℓ_G for the $\big(\mathcal{B}, H_G^*, H_G^*(\text{pt})\big)$-length where H_G^* denotes Borel cohomology with rational coefficients. Then Proposition 5.15 can be generalized to

$$\ell(X) \le \ell_G(X) \cdot |W| \cdot |\Gamma| \qquad \text{for any } G\text{-space } X$$

where $\Gamma = G/G_0$. Thus the length of X as a G-space and the length of X as a T-space are related in a way similar to Propositions 2.14 and 2.17.

   b) Observe that the upper bound $|W| \cdot |\Gamma|$ for $\ell(G/H)$ can be much larger than $1 + \dim(G/H)/T \le 1 + \dim G - \dim T$. For example, if $G = U(n)$ then $W = S_n$ is the symmetric group, hence $|W| = n!$, whereas $\dim U(n) = n^2$ and $\dim T = n$; similarly for $SU(n)$, $O(n)$ or $SO(n)$. On the other hand, in order to estimate $\mathcal{A}\text{-cat}(X)$ or $\mathcal{A}\text{-genus}(X)$ from above by $1 + \dim(X/G)$ we had to assume that $\mathcal{A}$ contains $\widehat{\mathcal{A}}_X$ which consists of all finite disjoint unions of orbits of X. If $\ell = (\mathcal{A}, h^*, I)$-length satisfies the strong normalization property as stated in 4.16 then $\ell = (\widehat{\mathcal{A}}, h^*, I)$-length, hence

$$\ell(X) \le \widehat{\mathcal{A}}\text{-cat}(X) \le 1 + \dim(X/G)$$

provided $\mathcal{A}_X \subset \mathcal{A}$. This applies to the lengths defined in 4.3–4.5 for $\mathbb{Z}/p$ or S^1. It is obviously false if ℓ does not satisfy the strong normalization property, for instance if G is a $(p\text{-})$torus of rank at least 2.

Proof of Proposition 5.15:
   In the case $\operatorname{rk}(H) = \operatorname{rk}(G)$ there exists $g \in G$ with $T \subset gHg^{-1}$. Thus $gH \in G/H$ is fixed by the action of T, so $\ell(G/H) = \ell(\text{pt}) = \infty$.
   Now assume that $\operatorname{rk}(H) < \operatorname{rk}(G) =: r$. We distinguish two cases depending on whether G is connected or not. Suppose first that G is connected. We want to show that $\ell(G/H) \le |W|$. We may assume that $H \cap T$ contains a maximal torus of H. If not we replace H by a conjugate group $H^g := gHg^{-1}$ such that $H^g \cap T$ contains a maximal torus of H^g. Since G/H and G/H^g are T-homeomorphic (they are even homeomorphic as G-spaces) we have $\ell(G/H^g) = \ell(G/H)$.
   The maps $BH \to BG$ and $BT \to BG$ turn both $H^*(BH)$ and $H^*(BT)$ into $H^*(BG)$-modules. The homomorphism $H^*(BG) \longrightarrow H^*(BT)$ is injective and maps $H^*(BG)$ onto the set of all elements of $H^*(BT)$ fixed by the action of W on $H^*(BT)$. Moreover,

$$H_T^*(G/H) \cong H^*(BH) \otimes_{H^*(BG)} H^*(BT).$$

See [Hs], §III.1, Proposition 1, for these results. Choose a subgroup H_1 of T of codimension 1 and containing $H \cap T$. There exist integers $a_1, \ldots, a_r$ such that

$$H_1 = \{(z_1, \ldots, z_r) \in T \colon \prod_{i=1}^{r} z_i^{a_i} = 1\}.$$

Then $I_1 := \ker\left(H_T^*(\mathrm{pt}) \cong H^*(BT) \longrightarrow H^*(BH_1) \cong H_T^*(T/H_1)\right)$ is the ideal generated by the polynomial

$$\pi_1 := \sum_{i=1}^{r} a_i c_i \in \mathbb{Q}[c_1, \ldots, c_r] \cong H_T^*(\mathrm{pt}).$$

Now let $w_1 = e, w_2, \ldots, w_k$ be the elements of the Weyl group W. W acts on T via conjugation and we write z^w for the action of $w \in W$ on $z \in T$. Thus if $w = gT$ for some $g \in NT$ then $z^w = gzg^{-1}$. Similarly we write ω^w for the action of $w \in W$ on $\omega \in H_T^*(\mathrm{pt}) \cong H^*(BT)$. Setting $H_\kappa := H_1^{w_\kappa}$ and $K_\kappa := \ker\left(H_T^*(\mathrm{pt}) \longrightarrow H_T^*(T/H_\kappa)\right)$ we obtain $K_\kappa = (\pi_\kappa)$ with $\pi_\kappa := \pi_1^{w_\kappa}$. In order to prove $\ell(G/H) \le k = |W|$ it suffices to show that for each $\omega_\kappa \in K_\kappa$, $\kappa = 1, \ldots, k$, the product $\omega := \omega_1 \cdot \ldots \cdot \omega_k$ maps to zero in $H_T^*(G/H)$. Now ω is a multiple of the product $\pi := \pi_1 \cdot \ldots \cdot \pi_k$ because π_κ divides ω_κ. We shall show that π annihilates $H_T^*(G/H)$, that is $\pi \cdot (1_H \otimes 1_T) = 1_H \otimes \pi$ is zero in $H_T^*(G/H) \cong H^*(BH) \otimes H^*(BT)$. Obviously π is invariant under the action of W on $H^*(BT)$, hence π lies in $H^*(BG) \cong H^*(BT)^W$. Therefore we may write $1_H \otimes \pi = (\pi \cdot 1_H) \otimes 1_T$ and it suffices to show $\pi \cdot 1_H = 0$ in $H^*(BH)$. Now $\pi \cdot 1_H = 0$ iff $\pi \cdot 1_{H \cap T} = 0 \in H^*\left(B(H \cap T)\right)$ because $H \cap T$ contains a maximal torus of H so that the homomorphism $H^*(BH) \to H^*\left(B(H \cap T)\right)$ is injective. Since $H \cap T \subset H_1$ we need only show $\pi \cdot 1_{H_1} = 0 \in H^*(BH_1)$. By construction, $\pi \cdot 1_{H_1}$ is a multiple of $\pi_1 \cdot 1_{H_1}$. And $\pi_1 \cdot 1_{H_1}$ is zero because π_1 lies in the kernel of the restriction homomorphism $H^*(BT) \to H^*(BH_1)$. This implies $\pi \cdot 1_H = 0$ as claimed.

It remains to consider the case where G is not connected. Since the component G_0 of the unit element of G is a normal subgroup, HG_0 is a (closed) subgroup of G. The inclusion $H \hookrightarrow HG_0$ induces a G-map $\pi\colon G/H \longrightarrow G/HG_0$. The homogeneous space G/HG_0 is isomorphic to the factor space of $\Gamma = G/G_0$ by HG_0/G_0. In particular, $G/HG_0 = \{g_1 HG_0, \ldots, g_n HG_0\}$ is finite with $n = |G/HG_0| \le |\Gamma|$. Moreover,

$$\pi^{-1}(g_\nu HG_0) = g_\nu HG_0/H = g_\nu G_0 H/H \subset G/H.$$

Next one checks that the G_0-map

$$G_0 \to g_\nu G_0 H/H, \quad g \mapsto gg_\nu H = g_\nu(g_\nu^{-1}gg_\nu)H,$$

is well defined and surjective. Therefore $g_\nu G_0 H/H \cong G_0/(G_0 \cap g_\nu Hg_\nu^{-1})$ is a homogeneous G_0-space. We know already that $\ell(g_\nu G_0 H/H) \le |W|$ according to the first case, hence we obtain

$$\ell(G/H) = \ell\big(\coprod_{\nu=1}^{n} g_\nu G_0 H/H\big) \le \sum_{\nu=1}^{n} \ell(g_\nu G_0 H/H) \le |G/HG_0| \cdot |W|.$$

$\square$

Chapter 6

The length and Conley index theory

6.1 Introduction

Starting with this chapter we shift our attention from computing the category or genus with the help of the length to using the length directly in critical point theory. By critical point theory we really mean the study of gradient-like flows. In many applications one is not only interested in the stationary points of the flow (critical points of the potential) but also in other flow orbits like connecting (heteroclinic) orbits.

A very elegant and useful tool in the study of qualitative properties of non-linear dynamical systems is the Conley index. Generalizing the Morse index of a non-degenerate critical point of a differentiable function it associates to an isolated invariant set of a flow a space with base point (or rather the homotopy type of a space with base point). Given a flow on a space X Conley index (or Morse-Conley) theory relates the topology of X and the behavior of the flow in a neighborhood of isolated invariant sets (Morse sets of a Morse decomposition) in a way similar to Morse theory. Typically one obtains results about the existence of isolated invariant sets and connecting orbits between them.

The geometric part of Conley index theory carries over to the equivariant setting without changes. In order to apply the Conley index one has to use methods from algebraic topology. We want to show that the length is a useful tool in this respect. One should mention that it is not clear how to generalize important concepts like connection matrices or transition matrices to the equivariant setting. This is due to the fact that all non-equivariant (co)homology groups are modules over $H^*(\mathrm{pt};\mathbb{F}) \cong H_*(\mathrm{pt};\mathbb{F}) \cong \mathbb{F}$, hence they are vector spaces if $\mathbb{F}$ is a field. If the flow is invariant under the action of a compact Lie group G then one should work with equivariant cohomology theories, for instance with Borel cohomology H_G^*. Then the cohomology groups are modules over $R = H_G^*(\mathrm{pt};\mathbb{F})$, in particular they are usually not free R-modules which causes various problems. It is interesting that the length is useful precisely because the cohomology modules are not free.

We shall restrict ourselves to the basic version of the Conley index as developed in [Con], [ConZ] or [Sa], that is to the study of flows on locally compact metric spaces. This suffices for the applications in chapters 8 and 9 because there we study bifurcation problems where a reduction to a finite-dimensional situation can be performed (via Lyapunov-Schmidt or the center manifold theorem). Extras can be included at additional costs. For example, it is possible to study semiflows on infinite-dimensional spaces which satisfy some compactness condition similar to the Palais-Smale condition as in [Ry]. We leave this kind of generalization to the future.

We proceed as follows. In §6.2 we recall the basic notions related to the equivariant Conley index. In §6.3 we use the length to study isolated invariant sets with a fixed Morse decomposition. Finally in §6.4 we study the behavior of the length under continuation.

6.2 The equivariant Conley index

In this section we fix notation and collect those results about equivariant flows and Conley index theory which are known in the non-equivariant version and translate immediately to our situation. General references are the papers by Conley [Con], Conley-Zehnder [ConZ] and Salamon [Sa].

Let X be a locally compact metric space. By a (local) flow on X we mean a continuous map $\varphi: D \to X$ defined on an open subset D of $X \times \mathbb{R}$ with $D \cap \{x\} \times \mathbb{R}$ an interval containing $(x, 0)$. The maps $\varphi^t = \varphi(-, t)$, $t \in \mathbb{R}$, satisfy $\varphi^0 = \mathrm{id}$ and $\varphi^{t+s} = \varphi^t \circ \varphi^s$. We assume that a compact Lie group G acts on X and that the maps φ^t are equivariant. φ is said to be *gradient-like* if there exists a continuous, G-invariant *Lyapunov function* $L: X \to \mathbb{R}$, that is $L(\varphi(x,t)) < L(x)$ for any $x \in X$ and $t > 0$ with $\varphi(x,t) \neq x$. A subset S of X is called *invariant* if it is both G-invariant and flow-invariant ($\varphi^t(S) \subset S$ for all t). Given a G-invariant subset Y of X we write

$$\mathrm{inv}(Y) = \mathrm{inv}(Y, \varphi) := \{x \in X : \varphi(x, t) \in Y \text{ for all } t\}$$

for the *maximal invariant subset* of Y. The α- and ω-*limit sets* of Y are defined as

$$\alpha(Y) := \mathrm{inv}\left(\mathrm{clos}\left(\varphi(Y \times \mathbb{R}^-)\right)\right)$$

and

$$\omega(Y) := \mathrm{inv}\left(\mathrm{clos}\left(\varphi(Y \times \mathbb{R}^+)\right)\right).$$

If x is a non-stationary point of φ then we call the flow orbit $\varphi^t(x)$ of x a *connecting orbit* from $\alpha(x)$ to $\omega(x)$. A G-*Morse decomposition* of a compact invariant set $S \subset X$ is a finite family $(M(\pi): \pi \in P)$ of pairwise disjoint compact invariant sets $M(\pi) \subset S$ with the following property:

> There exists an ordering $\pi_1, \ldots, \pi_n$ of P such that for every $x \in S - \bigcup_{\pi \in P} M(\pi)$ there exist indices $i, j \in \{1, \ldots, n\}$ with $i < j$ and $\omega(x) \subset M(\pi_i)$ and $\alpha(x) \subset M(\pi_j)$.

A compact invariant set $A \subset S$ is called an *attractor* if there exists a neighborhood U of A in S with $A = \omega(U)$. Similarly, a compact invariant set $A^* \subset S$ is called a *repeller* if there exists a neighborhood U of A^* in S with $A^* = \alpha(U)$. A G-invariant *attractor-repeller pair* of S is a G-Morse decomposition (A, A^*) such that A is an attractor and A^* is a repeller.

A compact invariant set $S \subset X$ is called *isolated* if there exists a G-neighborhood U of S in X such that $S = \mathrm{inv}(U)$. U is then called *isolating*. The basic notion of Conley index theory is that of an index pair for an isolated invariant set.

Index pairs:

Let S be an isolated invariant set. A pair (N_1, N_0) of compact G-invariant subsets $N_0 \subset N_1$ of X is a *G-index pair for S in X* if

— $N_1 - N_0$ is an isolating neighbourhood of S in X;

— N_0 is *positively invariant* with respect to N_1, that is if $x \in N_0$ and $\varphi(x, s) \in N_1$ for all $0 \le s \le t$ then $\varphi(x, s) \in N_0$ for $0 \le s \le t$;

— N_0 is an *exit set for N_1*, that is if $x \in N_1$ and $\varphi(x, t) \notin N_1$ for some $t > 0$ then there exists $s \in [0, t]$ with $\varphi(x, s) \in N_0$.

A G-index pair (N_1, N_0) is said to be *regular* if the function $\tau \colon N_1 \to [0, \infty]$ defined by

$$\tau(x) = \begin{cases} \sup\{t > 0 \colon \varphi(\{x\} \times [0, t]) \subset N_1 - N_0\} & \text{if } x \in N_1 - N_0; \\ 0 & \text{if } x \in N_0; \end{cases}$$

is continuous. Since τ is obviously G-invariant it induces an equivariant map

$$\tau^{-1}[0, \infty) \longrightarrow N_0, \ x \mapsto \varphi\big(x, \tau(x)\big).$$

If S is a repeller then $\tau^{-1}[0, \infty) = N_1 - S$.

Given an isolated invariant set $S \subset X$ and an isolating G-neighbourhood U of S there exists a G-index pair (N_1, N_0) contained in U. Furthermore for any G-neighbourhood U of N_0 in N_1 there exists a G-invariant subset N_0' of U containing N_0 and such that (N_1, N_0') is a regular G-index pair for S.

Conley-index:

If (N_1, N_0) and (N_1', N_0') are two G-index pairs for S then the quotient spaces N_1/N_0 and N_1'/N_0' are G-homotopy equivalent respecting the base points. The *Conley index* $C(S) = C(S, \varphi)$ of S is defined as the based G-homotopy type of N_1/N_0. We are using the convention that $N_1/\emptyset = N_1 \sqcup \mathrm{pt}$. We shall sometimes write $h^*\big(C(S)\big)$ for $h^*(N_1, N_0) = h^*(N_1/N_0, \mathrm{pt})$.

The applicability of the Conley index rests on the fact that it is invariant under certain deformations.

Parametrized flows and continuation:

Let Λ be a topological space with trivial G-action. We assume that Λ is connected, locally compact and locally contractible as in [Sa], although this can sometimes be relaxed. For our purposes, $\Lambda = [0, 1]$ is a good example. An *equivariant product flow* parametrized by Λ is an equivariant local flow φ on $\Lambda \times X$ such that $\varphi(\lambda, x, t) = \big(\lambda, \varphi_\lambda(x, t)\big)$. So every φ_λ is an equivariant local flow on X. To formulate the *continuation property* of the Conley index we consider the set

$$\mathcal{S} = \{(\lambda, S) \colon \lambda \in \Lambda \text{ and } S \text{ is an isolated invariant set of } \varphi_\lambda\}.$$

Consider $(\lambda_0, S_0) \in \mathcal{S}$ and a compact isolating G-neighbourhood N of S_0 in X (for the flow φ_{λ_0}). Then the set

$$\Lambda_N = \{\lambda \in \Lambda \colon N \text{ is isolating for } \varphi_\lambda\}$$

is an open neighborhood of λ_0 in Λ. Consider the section

$$\sigma_N : \Lambda_N \longrightarrow S, \quad \lambda \mapsto \big(\lambda, \mathrm{inv}(N, \varphi_\lambda)\big).$$

We give S the finest topology such that all σ_N are continuous. A basis of the topology of S consists of all $\sigma_N(K)$ with N as above and $K \subset \Lambda(N)$ open. Observe that σ_N is injective. Therefore σ_N is a local homeomorphism and so is the natural projection $S \to \Lambda$. Finally, a section $\sigma : K \to S$ defined on a subset K of Λ is continuous if and only if

$$S_K = \{(\lambda, x) : \lambda \in K \text{ and } x \in \mathrm{inv}(N, \varphi_\lambda)\}$$

is an isolated invariant set for the flow φ restricted to $K \times X$. If (λ, S) and (λ', S') lie in the same component of S — (λ, S) and (λ', S') are then said to be *related by continuation* — then their Conley indices are equal: $C(S, \varphi_\lambda) = C(S', \varphi_{\lambda'})$. This is the important continuation invariance of the Conley index.

6.3 Stationary solutions and connecting orbits

Let the compact Lie group G act on the locally compact, metric space X and let φ be a local equivariant flow on X. We fix a set $\mathcal{A}$ of G-spaces, a continuous multiplicative G-cohomology theory h^* with coefficient ring $R = h^*(\mathrm{pt})$ and a noetherian ideal $I \subset R$. As usual we write ℓ for $(\mathcal{A}, h^*, I)$-length.

6.1 Theorem:
Consider a compact invariant set $S \subset X$ of φ and a G-Morse decomposition $(M_1, \ldots, M_r)$ of S. Then

$$\ell(S) \leq \ell(M_1 * \ldots * M_r) \leq \sum_{i=1}^{r} \ell(M_i)$$

where $*$ denotes the join (see §2.3).

Proof:
The second inequality is a consequence of Corollary 4.10. To prove the first inequality let $J_r S$ denote the join of r copies of S. Then $M := M_1 * \ldots * M_r$ is a closed G-subset of $J_r S$. Because of the continuity property (see Theorem 4.7) there exists an open G-neighborhood U of M in $J_r S$ with $\ell(U) = \ell(M)$. Since M is compact there exist open G-neighborhoods U_i of M_i in S such that $U_1 * \ldots * U_r \subset U$. And since

$$\ell(M) \leq \ell(U_1 * \ldots * U_r) \leq \ell(U) = \ell(M)$$

we may assume $U = U_1 * \ldots * U_r$. We now construct a G-map $f : S \to U$ which implies $\ell(S) \leq \ell(U) = \ell(M)$ because of the monotonicity of ℓ. We need the following lemma.

6.2 Lemma:

Let φ be an equivariant flow on the compact G-space S and let (A, A^*) be a G-invariant attractor-repeller pair of S. Then for any G-neighborhood U of A there exists an invariant map $T: S - A^* \longrightarrow [0, \infty)$ such that $\varphi(x, T(x)) \in U$ for all $x \in S - A^*$.

Proof:

We may assume that U is positively invariant. Choose a regular G-index pair (N_1, N_0) for A^*. Thus we have an invariant map $\tau: N_1 - A^* \longrightarrow [0, \infty)$ with $\varphi(x, \tau(x)) \in N_0$ and $\tau(x) = 0$ for $x \in N_0$. By compactness there exists $T_0 \geq 0$ such that for all $x \in S - \text{int}(N_1 - N_0) \subset S - A^*$: $\varphi(x, T_0) \in U$. The map

$$T: S - A^* \longrightarrow [0, \infty), \quad T(x) = \begin{cases} T_0 + \tau(x) & \text{if } x \in N_1 - A^* \\ T_0 & \text{if } x \in S - N_1 \end{cases}$$

is the desired map.   □

To finish the proof of Theorem 6.1 we set

$$S_i := \{x \in S : \omega(x) \subset M_i \cup M_{i+1} \cup \ldots \cup M_r\}$$

for $i = 1, \ldots, r+1$. Then $S_1 = S$, $S_{r+1} = \emptyset$ and (M_i, S_{i+1}) is an attractor-repeller pair of S_i. Lemma 6.2 yields invariant maps $T_i: S_i - S_{i+1} \longrightarrow [0, \infty)$, $i = 1, \ldots, r$, such that $\varphi(x, T_i(x)) \in U_i$. Now $S_i - S_{i+1}$ is a closed G-subset of the normal G-space $S - S_{i+1}$. Hence, T_i can be extended to $S - S_{i+1}$ as a G-invariant map (cf. the Tietze-Gleason theorem 2.1). Since U_i is open the set

$$V_i = \{x \in S - S_{i+1} : \varphi(x, T_i(x)) \in U_i\}$$

is an open G-neighborhood of $S_i - S_{i+1}$ in $S - S_{i+1}$, hence in S. Now choose a G-invariant partition of unity $\pi = \{\pi_1, \ldots, \pi_r\}$ subordinated to $\{V_1, \ldots, V_r\}$ and define

$$f(x) := [\pi_1(x), \varphi(x, T_1(x)), \ldots, \pi_r(x), \varphi(x, T_r(x))].$$

The map $f: S = \bigcup_{i=1}^{r} V_i \longrightarrow U = U_1 * \ldots * U_r$ is well defined and continuous because $T_i(x)$ need only be defined for $x \in V_i$. It is obviously equivariant.   □

Theorem 6.1 is a generalization of the fact that any G-invariant C^1-function $F: S \to \mathbb{R}$ defined on a compact G-manifold S has at least $\ell(S)$ G-orbits of critical points if every critical G-orbit can be mapped equivariantly into an element of $\mathcal{A}$; see Theorem 2.19. The left inequality in 6.1 is useful for example in the case $G = \mathbb{Z}/p$. Namely, if $\mathbb{Z}/p$ acts freely on a sphere $S^{2n-1} = S$ then $\ell_0(S) = n$ and $\ell_1(S) = n+1$. So at first sight we get only about half of the critical G-orbits of $F: S \to \mathbb{R}$ which are known to exist. We obtain the correct result using the fact that $\ell_0(J_r G) = [(r+1)/2]$ and $\ell_1(J_r G) = [(r+2)/2]$; see Remark 5.4b). In this case it is also possible to work with the "length" ℓ defined in Remark 4.14. By the way, the factor $1/2$ occurs relatively often in $\mathbb{Z}/p$-equivariant problems when one is forced to use some kind of index theory instead of the Lusternik-Schnirelmann category; see for example [Ta], [MiT] or [BC1], Remark 3.5b). In §7.5 we shall get rid of this factor for the bifurcation problem of [BC1]. This can also be achieved in other situations.

The next result shows that under additional assumptions one can use the knowledge of $\ell(S)$ to analyze the flow on S.

6.3 Theorem:

Suppose ℓ satisfies the strong normalization property (see Remark 4.16). Let $S \subset X$ be a compact invariant set such that the flow φ on S is gradient-like. Assume moreover that for any stationary G-orbit Gx on S there exists $A(x) \in \mathcal{A}$ and a G-map $Gx \to A(x)$. Then there exist at least $k = \ell(S)$ different G-orbits of stationary solutions on S. If there are only finitely many stationary G-orbits on S then there exist k such orbits Gx_i and connecting orbits from Gx_{i+1} to Gx_i, $i = 1, \ldots, k-1$.

Proof:

We may assume that there are only finitely many G-orbits $(Gx_\pi : \pi \in P)$ of stationary solutions on S. Since the flow is gradient-like this family is a G-Morse decomposition of S. Theorem 6.1 gives

$$k = \ell(S) \leq \sum_{\pi \in P} \ell(Gx_\pi) = |P|.$$

So far we have not used the strong normalization property. The flow induces a partial order on P: $\pi < \pi'$ if and only if there exist sequences $\pi_0 = \pi, \pi_1, \ldots, \pi_r = \pi'$ in P and $y_1, \ldots, y_r \in S$ such that $\alpha(y_j) \subset Gx_{\pi_j}$ and $\omega(y_j) \subset Gx_{\pi_{j-1}}$ for $j = 1, \ldots, r$. A subset $J \subset P$ is an interval if $\pi < \pi' < \pi''$ and $\pi, \pi'' \in J$ imply $\pi' \in J$. For any interval J we set

$$M(J) := \{x \in S : \alpha(x) \cup \omega(x) \subset \bigcup_{\pi \in J} Gx_\pi\}.$$

$M(J)$ is a compact invariant set (see [Sa], Proposition 3.4). Let $\mathcal{J}$ be the set of all totally ordered intervals J of P. We have to show that there exists a $J \in \mathcal{J}$ with $|J| \geq k$. Set $r := \max\{|J| : J \in \mathcal{J}\}$. We decompose P into disjoint subsets $P_1, \ldots, P_r$ such that P_1 consists of the minimal elements of P and P_{i+1} consists of the minimal elements of $P - \bigcup_{j=1}^{i} P_j$. For each $\pi \in P_i$ there exist elements $\pi_j \in P_j$, $1 \leq j \leq i-1$, such that $\pi_1 < \pi_2 < \ldots < \pi_{i-1} < \pi$. For $i = 1, \ldots, r$ we set $M_i := \bigcup_{\pi \in P_i} Gx_\pi$. The family $M_1, \ldots, M_r$ is a G-Morse decomposition of S. Since each M_i consists of finitely many G-orbits there exists a G-neighborhood U_i of M_i in S and a G-retraction $U_i \to M_i$. Here we also use that S is completely regular as a compact space. In the proof of Theorem 6.1 we constructed a G-map $S \to U_1 * \ldots * U_r$. Using this map and the retractions we obtain a G-map $S \to M_1 * \ldots * M_r$. Therefore

$$k = \ell(S) \leq \sum_{i=1}^{r} \ell(M_i) = r$$

because by our assumptions on ℓ and $\mathcal{A}$ we know $\ell(M_i) = 1$.   □

6.4 Remark:

a) Recall from 4.15 that ℓ satisfies the strong normalization property if $G = \mathbb{Z}/p$ or if $G = S^1 \times \Gamma$, where Γ is any finite group, using the lengths defined in chapter 4. If p is an odd prime then Theorem 6.3 is also true for $\ell := \ell_0 + \ell_1 - 1$ as defined in Remark 4.14. To see this observe that in this case we have $\mathcal{A} = \{G\}$, so by assumption G acts freely on S. The G-map $S \to M_1 * \ldots * M_r$ constructed in the proof of Theorem 6.3 yields a G-map $S \to J_r G$ because each M_i can be mapped equivariantly into G (M_i consists only of finitely many G-orbits). Thus we also obtain $k = \ell(S) \leq \ell(J_r G) = r$; cf. Remark 5.4b).

b) Theorem 6.1 is also true if we replace the length by $\mathcal{A}$-genus. The properties listed in Proposition 2.15 suffice. For the category one can show that $\mathcal{A}\text{-cat}_S(S) \leq \sum_{i=1}^{r} \mathcal{A}\text{-cat}_S(M_i)$ provided S and all elements of $\mathcal{A}$ are G-ANRs. Theorem 6.3 continues to hold for $\mathcal{A}$-genus and for $\mathcal{A}$-category under additional assumptions, too. An interesting choice of $\mathcal{A}$ is $\widehat{\mathcal{A}}_S$ (see §2.4) because then $\mathcal{A}$-genus and $\mathcal{A}$-cat satisfy automatically the strong normalization property. And if G is a (p-)torus and S is G-homeomorphic to a sphere in a representation space of G then one can use Theorem 2.20 to compute $\mathcal{A}$-genus(S) and $\mathcal{A}$-cat(S).

c) One can analyze the flow on S further. For instance, if $G = \mathbb{Z}/2$ and if there are only finitely many G-orbits of stationary solutions on S then there exist k such orbits, $Gx_1, \ldots, Gx_k$, say, with $H^{i-1}(\mathcal{C}(x_i); \mathbb{F}_2) \neq 0$. This follows from the Morse inequalities if S is a manifold, the flow is a gradient flow and all stationary solutions are hyperbolic. Moreover, if in addition the flow is a Morse-Smale flow and there are precisely $k = \ell(S)$ stationary G-orbits on S then S is G-homeomorphic to the sphere S^{k-1}. This can be proved by induction on k. In the general case let $Gx_1, \ldots, Gx_r$ be the set of all stationary G-orbits so that the total order (given by the indices) extends the flow induced order. Consider the sets

$$\Sigma_i = \{x \in S \colon \alpha(x) \subset Gx_1 \cup \ldots \cup Gx_i\}$$

for $j = 0, \ldots, r$. Clearly $\Sigma_0 = \emptyset$, $\Sigma_r = S$ and (Σ_{i-1}, Gx_i) is an attractor-repeller pair of Σ_i. It is not difficult to show that $\ell(\Sigma_{i+1}) \leq \ell(\Sigma_i) + 1$. Fix some $i \in \{1, \ldots, k\}$ and let $j = j(i)$ be the smallest integer such that $\ell(\Sigma_j) = i$. Then we claim that $H^{i-1}(\mathcal{C}(x_j)) \neq 0$. To see this observe that $h^*(\mathcal{C}(Gx_j)) \cong h^*(\Sigma_j, \Sigma_{j-1})$ with $h^* = H_G^*(-; \mathbb{F}_2)$. For $w \in h^1(\text{pt})$ we know $w^{i-1} \cdot 1_{\Sigma_j} \neq 0 \in h^{i-1}(\Sigma_j)$ and $w^{i-1} \cdot 1_{\Sigma_{j-1}} = 0 \in h^{i-1}(\Sigma_{j-1})$. The long exact sequence of the pair (Σ_j, Σ_{j-1}) shows that $h^{i-1}(\Sigma_j, \Sigma_{j-1}) \neq 0$. Finally

$$h^{i-1}(\Sigma_j, \Sigma_{j-1}) \cong h^{i-1}(\Sigma_j/\Sigma_{j-1}, \text{pt})$$
$$\cong H_G^{i-1}(\mathcal{C}(x_j) \vee \mathcal{C}(-x_j), \text{pt}; \mathbb{F}_2)$$
$$\cong H^{i-1}(\mathcal{C}(x_j), \text{pt}; \mathbb{F}_2).$$

Here we wrote $\pm x_j$ for Gx_j. The last isomorphism comes from the fact that $G = \mathbb{Z}/2$ acts freely on $\mathcal{C}(x_j) \vee \mathcal{C}(-x_j) - \text{pt}$ and that without loss of generality the inclusion $\text{pt} \hookrightarrow \mathcal{C}(x_j) \vee \mathcal{C}(-x_j) \cong \mathcal{C}(\pm x_j)$ is a cofibration. This is the case if one defines the Conley index using a regular index pair. It is worth noting that the statements about the non-equivariant Conley indices $\mathcal{C}(x_i)$ continue to hold under non-equivariant perturbations of the flow. Thus the $2k$ stationary solutions

$\pm x_1, \dots, \pm x_k$ must pertain after perturbing the flow. They need not form G-orbits any more. It is also possible to deduce results on the existence of connecting orbits after a non-equivariant perturbation. We do not pursue this because this involves the (non-equivariant) connection matrices associated to the Morse decomposition $\pm x_1, \dots, \pm x_k$ of S; see [Fr].

d) It does not seem possible to obtain similar results for $G = \mathbb{Z}/2$ or $G = S^1$, say, using only the $\mathbb{F}$-vector space structure of $H_G^*(S; \mathbb{F})$ as codified for example in the connection matrix theory. Here one has to make additional non-degeneracy assumptions because for a degenerate isolated stationary solution x the critical vector space $H^*\big(\mathcal{C}(x)\big)$ can be very complicated. Of course, the connection matrix approach can yield the existence of other connecting orbits which cannot be detected with the help of the length.

6.4 Hyperbolicity and continuation

In this section we study the behavior of the length of isolated invariant sets under continuation. We fix a set $\mathcal{A}$ of G-spaces, a continuous multiplicative G-cohomology theory h^* and an ideal I of $R = h^*(\mathrm{pt})$. As usual we write ℓ for $(\mathcal{A}, h^*, I)$-length. Simple examples show that neither isolated invariant sets nor their lengths are invariant under continuation. Therefore we shall first relate the length of an isolated invariant set to the length of its Conley index. Consider an equivariant flow φ on a locally compact metric G-space X.

6.5 Definition:
An isolated invariant set S of φ is called h^*-*hyperbolic* if $h^*(S)$ and $h^*\big(\mathcal{C}(S)\big)$ are isomorphic R-modules.

Recall that we agreed to write $h^*\big(\mathcal{C}(S)\big)$ for $h^*(N_1/N_0, \mathrm{pt}) \cong h^*(N_1, N_0)$ if (N_1, N_0) is a G-index pair for S. Similarly we write $\ell\big(\mathcal{C}(S)\big)$ for $\ell(N_1/N_0, \mathrm{pt}) = \ell(N_1, N_0)$.

6.6 Proposition:
Let S be an isolated invariant set of φ. Then $\ell(S) \geq \ell\big(\mathcal{C}(S)\big)$ and equality holds if S is h^*-hyperbolic.

Proof:
Choose a G-index pair (N_1, N_0) of S such that $\ell(N_1) = \ell(S)$. This is possible because ℓ is continuous and because we may choose the index pair arbitrarily close to S. The inequality $\ell(N_1) \geq \ell(N_1, N_0)$ is obvious since $\gamma = 1_{N_1} \smile \gamma$ for any $\gamma \in h^*(N_1, N_0)$. So if an element of R annihilates 1_{N_1} then it annihilates $h^*(N_1, N_0)$. If S is h^*-hyperbolic then $\ell(S) = \ell\big(\mathcal{C}(S)\big)$ because the length depends only on the R-module structure. □

This simple result is useful because $\ell\big(\mathcal{C}(S)\big)$ is invariant under continuation. Let us comment a bit on the notion of h^*-hyperbolicity. Observe that $h^*(S)$ acts on $h^*\big(\mathcal{C}(S)\big)$ as follows. Given elements $\alpha \in h^*(S)$ and $\gamma \in h^*\big(\mathcal{C}(S)\big)$ there exists a G-neighborhood U of S such that α lies in the image of the restriction map $|S: h^*(U) \to h^*(S): \alpha = \beta|S$. This is due to the continuity of h^*. Let (N_1, N_0) be a G-index pair of S contained in U so that $\gamma \in h^*(N_1, N_0)$. Then we define

$$\alpha \cdot \gamma := (\beta|N_1) \smile \gamma \in h^*(N_1, N_0).$$

One checks easily that this definition is independent of the choice of (N_1, N_0). If U' is another G-neighborhood of S and $\beta' \in h^*(U')$ restricts to α, then there exists a G-neighborhood $U'' \subset U \cap U'$ with $\beta'|U'' = \beta|U''$. Now we may choose (N_1, N_0) contained in U'' whence it follows that

$$(\beta|N_1) \smile \gamma = (\beta'|N_1) \smile \gamma.$$

Thus $\alpha \cdot \gamma$ is well defined.

The isolated invariant set S is h^*-hyperbolic if there exists an element $u \in h^*\big(\mathcal{C}(S)\big)$ such that the homomorphism

$$h^*(S) \to h^*\big(\mathcal{C}(S)\big), \quad \alpha \mapsto \alpha \cdot u,$$

is an isomorphism. For example, if X is a manifold and S is a submanifold of stationary orbits of the flow such that the normal space N_x to S at a point x splits into a stable part N_x^s and an unstable part N_x^u then S is h^*-hyperbolic if the vector bundle N^u consisting of the unstable parts is orientable with respect to h^*. The element u can then be thought of being an orientation for this bundle because $h^*\big(\mathcal{C}(S)\big) \cong h^*(N^u, N^u - S)$. Here we identified S with the zero section of N^u.

In [Fl] Floer introduced the more special notion of $*$-hyperbolic sets. An isolated invariant set S is $*$-*hyperbolic* in the sense of Floer if it is H_G^*-hyperbolic in our sense and if it is an equivariant retract of some G-neighborhood. In [Fl] one can also find a criterion for an invariant submanifold S to be $*$-hyperbolic (hence H_G^*-hyperbolic) which generalizes our example above in that Floer allows a nontrivial flow on S (see [Fl], Proposition 1).

Now we consider the continuation problem for the length. Let Λ be a trivial G-space and φ be an equivariant product flow on $\Lambda \times X$ as in §6.2. We consider the set

$$\mathcal{S} := \{(\lambda, S): \lambda \in \Lambda \text{ and } S \subset X \text{ is an isolated invariant set of } \varphi_\lambda\}$$

with the topology defined in §6.2. For simplicity we write $\ell(\lambda, S)$ instead of $\ell(S, \varphi_\lambda)$.

6.7 Theorem:

The map $\ell: \mathcal{S} \to \mathbb{N}$, $(\lambda, S) \mapsto \ell(\lambda, S)$ *is upper-semi-continuous, i.e. for any* $(\lambda_0, S_0) \in \mathcal{S}$ *there exists a neighborhood* U *in* $\mathcal{S}$ *with* $\ell(\lambda, S) \leq \ell(\lambda_0, S_0)$ *for every* $(\lambda, S) \in U$. *Moreover,* ℓ *is continuous at an* h^*-*hyperbolic set* (λ_0, S_0).

Proof:

Given $(\lambda_0, S_0) \in S$ there exists an open G-neighborhood N of S_0 in X with $\ell(N) = \ell(S_0)$. Then for any $\lambda \in \Lambda_N = \{\lambda \in \Lambda : N$ is isolating for $\varphi_\lambda\}$:

$$\ell\big(\sigma_N(\lambda)\big) = \ell\big(\lambda, \text{inv}(N, \varphi_\lambda)\big) \leq \ell(N) = \ell(S).$$

Thus we may take $U = \sigma_N(\Lambda_N)$. Finally, if (λ_0, S_0) is h^*-hyperbolic then let K be the path-component of λ_0 in Λ_N. This is a neighborhood of λ_0 because we assumed Λ to be locally path-connected (even locally contractible). For any $\lambda \in K$ $\sigma_N(\lambda)$ is related by continuation to $\sigma_N(\lambda_0)$. Hence, using Proposition 6.6 and the continuation invariance of the Conley index we obtain

$$\ell\big(\sigma_N(\lambda)\big) \geq \ell\big(\mathcal{C}(\sigma_N(\lambda))\big) = \ell\big(\mathcal{C}(\sigma_N(\lambda_0))\big) = \ell\big(\sigma_N(\lambda_0)\big).$$

Thus ℓ must be constant in the neighborhood $\sigma_N(K)$ of (λ_0, S_0) as claimed. $\qquad\square$

6.8 Corollary:

If (λ, S) and (λ', S') are related by continuation and if (λ, S) is h^*-hyperbolic then $\ell(\lambda', S') \geq \ell(\lambda, S)$. If also (λ', S') is h^*-hyperbolic then $\ell(\lambda', S') = \ell(\lambda, S)$.

Chapter 7

The exit-length

7.1 Introduction

Given an isolated invariant set S of an equivariant gradient-like flow on X the knowledge of the category, genus or length of S yields estimates for the number of stationary flow orbits and connecting orbits on S. In the first chapters we assumed that S is a given manifold, for example a representation sphere, and tried to compute its category, genus or length. In this chapter we are interested in a situation where S itself is to be determined. More precisely, we study a bifurcation problem where S is a set of orbits which bifurcate from a given branch of stationary orbits. A priori we do not know anything about S at all.

To illustrate the problem consider a family of gradient-like flows φ_λ on $\mathbb{R}^n$ continuously parametrized by $\lambda \in \mathbb{R}$. We assume that a compact Lie group G acts orthogonally on $\mathbb{R}^n$ and that each φ_λ is equivariant. Suppose $0 \in \mathbb{R}^n$ is a stationary solution of φ_λ for every λ (this is automatic if the action is fixed point free), and that it is isolated for every λ different from a possible bifurcation point λ_0. Thus the Conley indices $\mathcal{C}(0, \varphi_\lambda)$ are well defined for $\lambda \neq \lambda_0$. If $\mathcal{C}(0, \varphi_\lambda)$ changes as λ passes λ_0 then bifurcation must occur. How can one measure this change and the size of the bifurcating set? Since $0 \in \mathbb{R}^n$ is fixed by G the length of $\mathcal{C}(0, \varphi_\lambda)$ is equal to $\ell(\mathrm{pt})$ independent of λ. It is natural to work with the length of the unstable set of 0 for φ_λ which we call $\ell^u(0, \varphi_\lambda)$. This set consists of all points (except 0) whose α-limit set is 0. An equivalent description of $\ell^u(0, \varphi_\lambda)$ is as the minimal length of an exit set of 0. Therefore we call it the exit-length of 0. If the origin is a hyperbolic stationary solution of φ_λ then this unstable set is homotopy equivalent to the sphere of the unstable tangent space at the origin. Thus at least for $(p\text{-})$tori the exit-length is closely related to the dimension of the unstable manifold. If $\ell^u(0, \varphi_\lambda)$ changes as λ passes λ_0 then the length of the bifurcating set should make up for the difference $|\ell^u(0, \varphi_{\lambda_0+\varepsilon}) - \ell^u(0, \varphi_{\lambda_0-\varepsilon})|$.

The main goal of this chapter is to prove such a result. As a corollary we obtain lower bounds for the number of bifurcating stationary solutions and connecting orbits. This provides an answer to the questions posed in 2.23 and improves and generalizes older bifurcation theorems as in [FaR1,2] (where G is either $\mathbb{Z}/2$ or S^1) or [BC1]. There the authors consider gradient flows and use minimax methods which only yield stationary flow orbits (critical points of the potential) but not connecting orbits. But even if one is only interested in the stationary solutions we improve the result of [BC1] for p-tori, for instance, by the factor 2 (if p is an odd prime). In addition, our approach works for very degenerate situations, i.e. we do not need any "generic" assumptions on the bifurcating sets.

The chapter is organized as follows. After the definition of the exit-length in §7.2 we study its behavior under continuation in §7.3. Then in §7.4 we relate the exit-length and the dual concept of entry-length. Moreover we prove a theorem on the lengths and exit-lengths of the Morse sets in a given Morse decomposition of S. Although it would be interesting to develop the theory in greater detail (and relate it for example to the connection matrix approach in Conley index theory) we concentrate on those results needed for the bifurcation theorems in §7.5. In this chapter it seems to be important to work with the length. Whether or not it is possible to use the genus or a version of the category is a difficult question.

Throughout this chapter we fix a set $\mathcal{A}$ of G-spaces, a continuous, multiplicative equivariant cohomology theory h^* and a noetherian ideal I of the coefficient ring $R = h^*(\text{pt})$. As usual we write ℓ for $(\mathcal{A}, h^*, I)$-length.

7.2 The exit-length of isolated invariant sets

As in Chapter 6 we study an equivariant local flow φ on a locally compact metric G-space X with the help of the length $\ell = (\mathcal{A}, h^*, I)$-length. Consider an isolated invariant set S and an isolating neighborhood N of S. The *unstable set* of S in N is by definition

$$\begin{aligned} N^u &= \{x \in N - S : \varphi^t(x) \in N \text{ for all } t \le 0\} \\ &= \{x \in N - S : \alpha(x) \subset S\}. \end{aligned}$$

Similarly, the *stable set* of S in N is

$$\begin{aligned} N^s &= \{x \in N - S : \varphi^t(x) \in N \text{ for all } t \ge 0\} \\ &= \{x \in N - S : \omega(x) \subset S\}. \end{aligned}$$

If M is another isolating neighborhood of S then there exists a real number $t \le 0$ such that

$$\varphi^t(N^u) \subset M^u \quad \text{and} \quad \varphi^t(M^u) \subset N^u.$$

Similarly, there exists $t \ge 0$ with

$$\varphi^t(N^s) \subset M^s \quad \text{and} \quad \varphi^t(M^s) \subset N^s.$$

Thus the lengths of N^u and N^s are independent of the choice of the isolating neighborhood N because of the monotonicity of ℓ.

7.1 Definition:
Given an isolated invariant set $S \subset X$ of φ the number $\ell^u(S) = \ell^u(S, \varphi) := \ell(N^u)$ is called the *exit-length* of S. Here N is an arbitrary isolating G-neighborhood of S. Similarly, the number $\ell^s(S) = \ell^s(S, \varphi) := \ell(N^s)$ is called the *entry-length* of S.

If X is a G-manifold and $S = \{x\} \subset X^G$ is a hyperbolic stationary orbit of φ then the exit-length of S is the length of the sphere of the unstable tangent space $T_x^u X$ of X at x: $\ell^u(S) = \ell(S(T_x^u X))$. For elementary abelian p-groups or tori this is closely related to the Morse index of x, at least if x is an isolated element of X^G. We now begin to study the exit-length.

7.2 Proposition:

For an isolated invariant set S we have $\ell^u(S) \leq \ell(S)$.

Proof:

By the continuity of ℓ we may choose the neighborhood N in the definition of the exit-length so small that

$$\ell(S) = \ell(N) \geq \ell(N^u) = \ell^u(S).$$

$\square$

A very useful description of the exit-length is contained in the next result.

7.3 Proposition:

a) *If (N_1, N_0) is a G-index pair of S then $\ell^u(S) \leq \ell(N_0)$.*

b) *For every G-neighborhood N of S there exists a G-index pair (N_1, N_0) contained in N with $\ell(N_0) = \ell^u(S)$.*

Consequently: $\ell^u(S) = \min\{\ell(N_0): (N_1, N_0) \text{ is a } G\text{-index pair of } S\}$.

Proof:

a) Given a G-index pair (N_1, N_0) of S we may assume that it is regular. If not we can replace N_0 by a bigger exit set N_0' contained in an arbitrarily small neighborhood of N_0 in N_1 and such that (N_1, N_0') is regular (see [Sa], 5.4). The continuity of ℓ implies $\ell(N_0') = \ell(N_0)$ if N_0' is close enough. For a regular G-index pair the map

$$\tau: N_1 \longrightarrow [0, \infty], \tau(x) := \begin{cases} \sup\{t \geq 0: \varphi(x, [0, t]) \subset N_1 - N_0\} & \text{if } x \in N_1 - N_0; \\ 0 & \text{if } x \in N_0; \end{cases}$$

is continuous. Thus we have an equivariant map $N_1^u \to N_0$, $x \mapsto \varphi\big(x, \tau(x)\big)$. The monotonicity of ℓ now gives $\ell^u(S) = \ell(N_1^u) \leq \ell(N_0)$.

b) Given N choose a regular G-index pair (M_1, M_0) contained in N. Let U be a closed neighborhood of $M_1^u \cap M_0$ in M_0 such that $\ell(U) = \ell(M_1^u \cap M_0)$. Since (M_1, M_0) is regular the map $\tau: M_1 \longrightarrow [0, \infty]$, defined as in a), is continuous. Now $\tau^{-1}(\infty) = S \cup M_1^s$ so τ induces a G-map

$$\psi: M_1 - (S \cup M_1^s) \to M_0, \quad x \mapsto \varphi\big(x, \tau(x)\big).$$

Moreover, $\psi(M_1^u) \subset M_0 \cap M_1^u$ and therefore $\ell^u(S) = \ell(M_1^u) = \ell(M_0 \cap M_1^u)$. By continuity we may choose a compact G-invariant neighborhood N_0 of $M_0 \cap M_1^u$ in M_0 with $\ell(N_0) = \ell^u(S)$. Now we define $N_1 := S \cup M_1^s \cup \psi^{-1}(N_0) \subset M_1$. This is a closed G-subset of M_1. It remains to prove that (N_1, N_0) is a G-index pair of S. First we check that $N_1 - N_0$ is a neighborhood of S. Suppose to the contrary that there exists a sequence $x_n \in M_1$ with $\text{dist}(x_n, S) \to 0$ as $n \to \infty$ and $x_n \notin N_1 - N_0$. Then $\tau(x_n) < \infty$, $\tau(x_n) \to \infty$ and $y_n := \varphi\big(x_n, \tau(x_n)\big) \in M_0 - N_0$. By compactness we may assume that $y_n \to y \in \text{clos}(M_0 - N_0) \subset M_0 - M_1^u$. Next observe that for every $t \leq 0$

$$\varphi(y, t) = \lim_{n \to \infty} \varphi(y_n, t) = \lim_{n \to \infty} \varphi\big(x_n, t + \tau(x_n)\big) \in M_1$$

because M_1 is closed and $\varphi\bigl(x_n, t + \tau(x_n)\bigr) \in M_1$ if $\tau(x_n) \geq -t$. It follows that $y \in M_1^u$, a contradiction. Thus $N_1 - N_0$ is a G-neighborhood of S. It is easy to see that this neighborhood is isolating, that N_0 is positively invariant with respect to N_1 and that N_0 is an exit set for N_1.   □

The results of this section continue to hold when one replaces the length by the genus because we only used the monotonicity and the continuity properties. If $G = \mathbb{Z}/p$ and p is an odd prime then we may also work with the "length" $\ell = \ell_0 + \ell_1 - 1$ as defined in Remark 4.14.

7.3 Continuation of the exit-length

In this section we want to study the behavior of the exit-length under continuation. Let φ be a product flow on $\Lambda \times X$ as in §6.2 and set

$$\mathcal{S} = \{(\lambda, S) : \lambda \in \Lambda \text{ and } S \text{ is an isolated invariant set of } \varphi_\lambda\}$$

with the topology defined in §6.2. Recall that the canonical projection $\mathcal{S} \to \Lambda$ is a local homeomorphism. For simplicity we write $\mathcal{C}(\lambda, S)$ respectively $\ell^u(\lambda, S)$ for $\mathcal{C}(S, \varphi_\lambda)$ respectively $\ell^u(S, \varphi_\lambda)$.

7.4 Theorem:

a) *The map $\ell^u \colon \mathcal{S} \longrightarrow \mathbb{N}$, $(\lambda, S) \mapsto \ell^u(\lambda, S)$, is upper-semi-continuous.*

b) *Consider a continuous section $\sigma \colon K \to \mathcal{S}$, $\sigma(\lambda) = (\lambda, S_\lambda)$, defined on a path-connected subset K of Λ. Suppose there exists $\kappa \in K$ such that S_κ is an invariant (with respect to φ_λ) subset of S_λ for all $\lambda \in K$. We do not assume that S_κ is isolated invariant with respect to the flow φ_λ for $\kappa \neq \lambda$. Then $\ell^u\bigl(\sigma(\kappa)\bigr) \leq \ell^u\bigl(\sigma(\lambda)\bigr)$ for every $\lambda \in K$. In particular, ℓ^u is continuous at $\sigma(\kappa)$.*

In order to prove Theorem 7.4 we need another cohomological description of the exit-length. The following result does not refer to the parametrized situation. It holds for all flows φ.

7.5 Theorem:

Let $S \subset X$ be an isolated invariant set of φ and (N_1, N_0) a G-index pair of S. The inclusion $i \colon S \hookrightarrow (N_1, N_0)$ induces a homomorphism $i^* \colon h^*(\mathcal{C}(S)) \cong h^*(N_1, N_0) \longrightarrow h^*(S)$. The exit-length $\ell^u(S)$ is the minimal number $k \geq 0$ such that there exist elements $A_1, \ldots, A_k \in \mathcal{A}$ with the following property.

(*) *For all $\omega_i \in I \cap \ker\bigl(R \to h^*(A_i)\bigr)$, $i = 1, \ldots, k$, the product $\omega_1 \cdot \ldots \cdot \omega_k \cdot 1_S \in h^*(S)$ is contained in the image of i^*. For $k = 0$ this means that 1_S is in the image of i^*.*

Consequently, if the inclusion $S \hookrightarrow N_1$ induces a monomorphism $h^*(N_1) \longrightarrow h^*(S)$ then $\ell^u(S) = \ell(N_0)$.

Proof:

First of all, observe that the map i^* is independent of (N_1, N_0). Namely, if (M_1, M_0) is another G-index pair of S then Salamon ([Sa], Lemma 4.7) constructs a G-homotopy equivalence $N_1/N_0 \longrightarrow M_1/M_0$ which makes the following diagram commutative.

$$
\begin{array}{ccc}
S & \longrightarrow & (N_1/N_0, \mathrm{pt}) \\
\downarrow {\scriptstyle \mathrm{id}} & & \downarrow \\
S & \longrightarrow & (M_1/M_0, \mathrm{pt})
\end{array}
$$

Because of Proposition 7.3 we may assume that $m := \ell(N_0) = \ell^u(S)$. Let $A_1, \ldots, A_m$ be elements of $\mathcal{A}$ such that

$$\omega_1 \cdot \ldots \cdot \omega_m \cdot 1_{N_0} = 0 \in h^*(N_0) \quad \text{for all } \omega_i \in I \cap \ker\left(R \to h^*(A_i)\right).$$

The long exact cohomology sequence of (N_1, N_0) implies that $\omega_1 \cdot \ldots \cdot \omega_m \cdot 1_{N_1} \in h^*(N_1)$ comes from $h^*(N_1, N_0)$. Therefore $\omega_1 \cdot \ldots \cdot \omega_m \cdot 1_S \in h^*(S)$ comes from $h^*(N_1, N_0)$ which implies that $\ell^u(S) = m$ satisfies $(*)$, hence, $\ell^u(S) \geq k$.

Next consider elements $A_1, \ldots, A_k$ of $\mathcal{A}$ satisfying $(*)$. Since I is noetherian there exist finitely many generators

$$\omega_{i,j} \in I \cap \ker\left(R \to h^*(A_i)\right), \quad i = 1, \ldots, k, \quad j = 1, \ldots, k_i.$$

For any k-tuple $J = (j_1, \ldots, j_k)$ of integers j_i satisfying $1 \leq j_i \leq k_i$ we consider the product $\omega_J := w_{1,j_1} \cdot \ldots \cdot w_{k,j_k} \in I$. Because of $(*)$ $\omega_J \cdot 1_S \in h^*(S)$ comes from an element $\alpha_J \in h^*\left(\mathcal{C}(S)\right) = h^*(N_1, N_0)$. α_J restricts to an element $\beta_J \in h^*(N_1)$ and $\beta_J - \omega_J \cdot 1_{N_1}$ maps to zero when restricted to $h^*(S)$. The continuity of h^* implies that there exists a G-neighborhood $N \subset N_1$ of S such that the finitely many elements $\beta_J - \omega_J \cdot 1_{N_1}$ map to zero in $h^*(N)$. Then for any G-index pair (M_1, M_0) contained in N and any k-tuple J as above $\omega_J \cdot 1_{M_1} \in h^*(M_1)$ comes from $\alpha_J \in h^*\left(\mathcal{C}(S)\right) = h^*(M_1, M_0)$. This implies that for any $\omega_i \in I \cap \ker\left(R \to h^*(A_i)\right)$, $i = 1, \ldots, k$, the product $\omega_1 \cdot \ldots \cdot \omega_k \cdot 1_{M_0} \in h^*(M_0)$ is zero since ω_i can be written as a linear combination of the $\omega_{i,j}$, $j = 1, \ldots, k_i$. Because of Proposition 7.3b) we may choose (M_1, M_0) such that $\ell(M_0) = \ell^u(S)$. Then we obtain $k \geq \ell(M_0) = \ell^u(S)$. $\square$

Proof of Theorem 7.4:

a) Given $(\kappa, S) \in \mathcal{S}$ with $k = \ell^u(\kappa, S)$, we choose elements $A_1, \ldots, A_k \in \mathcal{A}$ according to Theorem 7.5. As shown in the proof of 7.5 there exists an isolating (with respect to φ_κ) G-neighborhood N of S such that for each $\omega_i \in \ker\left(R \to h^*(A_i)\right)$ the product $\omega_1 \cdot \ldots \cdot \omega_k \cdot 1_N \in h^*(N)$ comes from $h^*\left(\mathcal{C}(\kappa, S)\right)$. Now, N is also isolating for φ_λ if λ is close to κ. Consider $\Lambda_N = \{\lambda \in \Lambda \colon N \text{ is isolating for } \varphi_\lambda\}$ and set $S_\lambda := \mathrm{inv}(N, \varphi_\lambda)$, $\sigma_N(\lambda) := (\lambda, S_\lambda)$ for $\lambda \in \Lambda_N$. We may assume that Λ_N is connected, so that $\sigma(\kappa)$ and $\sigma(\lambda)$ are related by continuation. Then $\omega_1 \cdot \ldots \cdot \omega_k \cdot 1_{S_\lambda} \in h^*(S_\lambda)$ comes from $h^*\left(\mathcal{C}(\sigma(\lambda))\right) \cong h^*\left(\mathcal{C}(\sigma(\kappa))\right)$, where the ω_i, $i = 1, \ldots, k$, are as above. Using 7.5 once more, we see that $\ell^u\left(\sigma(\lambda)\right) \leq k = \ell^u\left(\sigma(\kappa)\right)$ for all $\lambda \in \Lambda_N$.

b) Since $\sigma(\kappa)$ and $\sigma(\lambda)$ are related by continuation there exists a G-homotopy equivalence $f \colon \mathcal{C}\left(\sigma(\kappa)\right) \longrightarrow \mathcal{C}\left(\sigma(\lambda)\right)$. The construction of f as in [Sa], Section 6, makes the following diagram commute up to G-homotopy:

$$
\begin{array}{ccc}
S_\kappa & \longrightarrow & \mathcal{C}\big(\sigma(\kappa)\big) \\
\downarrow & & \downarrow{\scriptstyle f} \\
S_\lambda & \longrightarrow & \mathcal{C}\big(\sigma(\lambda)\big)
\end{array}
$$

Here and in the sequel we assume that a particular G-index pair (N_1, N_0) for S_κ has been chosen and we do not distinguish between N_1/N_0 and $\mathcal{C}(\kappa, S_\kappa)$; similarly for λ. All maps different from f in the above diagram are inclusions. For the commutativity of the diagram it is important that S_κ is an invariant subset of S_λ for all $\lambda \in K$. We obtain a commutative diagram on h^*-level:

$$
\begin{array}{ccc}
h^*\big(\mathcal{C}(\sigma(\lambda))\big) & \longrightarrow & h^*(S_\lambda) \\
\downarrow & & \downarrow \\
h^*\big(\mathcal{C}(\sigma(\kappa))\big) & \longrightarrow & h^*(S_\kappa)
\end{array}
$$

If $\omega \cdot 1_{S_\lambda} \in h^*(S_\lambda)$ comes from $h^*\big(\mathcal{C}(\sigma(\lambda))\big)$ for some $\omega \in R$ then $\omega \cdot 1_{S_\kappa}$ comes from $h^*\big(\mathcal{C}(\sigma(\kappa))\big)$. This implies $\ell^u\big(\sigma(\kappa)\big) \leq \ell^u\big(\sigma(\lambda)\big)$ using the description of the exit-length from Theorem 7.5. That ℓ^u is continuous at $\sigma(\kappa) = (\kappa, S_\kappa)$ follows from the upper-semi-continuity of ℓ and because σ is a homeomorphism onto its image.□

We do not know whether Theorem 7.4 continues to hold for the genus instead of the length (probably not). Our proof of 7.4 uses in an essential way the cohomological description of the exit-length in 7.5, hence it uses the definition of the length and not just its properties.

7.4 Properties of the exit-length

The results of this section will be needed for the proof of the bifurcation theorem in §7.5.

7.6 Proposition:
For any isolated invariant set S of X there exists a G-neighborhood N such that for all G-neighborhoods U of S in N: $\ell(\partial U) \leq \ell^u(S) + \ell^s(S)$.

Proof:
Because of Proposition 7.3 there exists a regular G-index pair (N_1, N_0) of S such that $\ell^u(S) = \ell(N_0)$. Similarly, let (M_1, M_0) be a regular G-index pair of S with respect to the inverse flow such that $\ell^s(S) = \ell(M_0)$. Set $N := N_1 \cap M_1$. We may assume that N is an isolating neighborhood of S. For any G-neighborhood U of S contained in N consider the sets

$$
A := \big\{x \in \partial U : \text{there exists } t \geq 0 \text{ with } \varphi(x, t) \notin N\big\}
$$

and

$$
B := \big\{x \in \partial U : \text{there exists } t \leq 0 \text{ with } \varphi(x, t) \notin N\big\}.
$$

Obviously, $\partial U = A \cup B$ since any orbit that stays inside N all the time is contained in S. Now the regularity of (N_1, N_0) implies the existence of a G-map $A \to N_0$. And the regularity of (M_1, M_0) gives a G-map $B \to M_0$. Thus using the subadditivity and the monotonicity of ℓ we obtain the desired inequality:

$$\ell(\partial U) \leq \ell(A) + \ell(B) \leq \ell(N_0) + \ell(M_0) = \ell^u(S) + \ell^s(S).$$

□

If $G = \mathbb{Z}/p$, p an odd prime, and $\ell = \ell_0 + \ell_1 - 1$ is defined as in Remark 4.14 then 7.6 is also true provided $\ell_1(\partial U) = \ell_0(\partial U) + 1$. This will be the case in the following proposition.

7.7 Proposition:

Let G be either equal to $\mathbb{Z}/p$, p a prime, or equal to $S^1 \times \Gamma$, where Γ is a finite group. Let $X \cong \mathbb{R}^n$ be a representation of G satisfying $X^G = 0$ (respectively $X^{S^1} = 0$ if $G = S^1 \times \Gamma$). Suppose $S = \{0\} \subset X$ is an isolated invariant set for the flow φ on X. Then $\ell^u(S) + \ell^s(S) = \ell(SX)$ where ℓ is as in Example 4.4 respectively Remarks 4.14 and 4.16; SX denotes the unit sphere of X.

Proof:

We only have to show $\ell^u(S) + \ell^s(S) \leq \ell(SX)$. Choose a regular G-index pair (N_1, N_0) of S and set $A := N_1^u \cap N_0$. Using the monotonicity of ℓ one sees as in the proof of Proposition 7.3 that $\ell^u(S) = \ell(A)$. Similarly $\ell^s(S) = \ell(B)$ with $B := M_1^s \cap M_0$ where (M_1, M_0) is a regular G-index pair for the inverse flow, i.e. M_0 is a regular entry set for φ. Next consider a small closed ball BX around 0 in X such that $A \cup B \subset X - BX$. There exists $t > 0$ such that $\varphi(A, -t)$ and $\varphi(B, t)$ are contained in the interior of BX. Setting $A' := \partial BX \cap \varphi(A \times [-t, 0])$ and $B' := \partial BX \cap \varphi(B \times [0, t])$ the piercing property 4.11 yields $\ell(A') \geq \ell(A)$ and $\ell(B') \geq \ell(B)$. Since $A' \cap B' = \emptyset$ the theorem 4.17 (respectively Proposition 4.18 for $G = \mathbb{Z}/p$) implies the desired inequality:

$$\ell^u(S) + \ell^s(S) = \ell(A) + \ell(B) \leq \ell(A') + \ell(B') \leq \ell(\partial BX) = \ell(SX).$$

□

Now we consider the following situation. We are given an isolated invariant set S and a G-Morse decomposition (M_1, M_2, M_3). We think of M_2 as a known solution and want to get information about M_1 and M_3. We shall encounter such a situation in a bifurcation setting where M_2 lies on a given branch of (stationary) orbits and M_1 and M_3 lie on the bifurcating branches. Suppose we know $\ell^u(S)$ and $\ell^u(M_2)$, and these exit-lengths differ. Then, obviously, S and M_2 are different and therefore $M_1 \cup M_3 \neq \emptyset$. Moreover it is a simple consequence of the properties of the length that $\ell(M_1) \geq \ell^s(M_1) \geq \ell^u(M_2) - \ell^u(S)$ because the unstable set of M_2 is contained in the union of the stable set of M_1 and the unstable set of S. Analogously we obtain $\ell(M_3) \geq \ell^u(M_3) \geq \ell^s(M_2) - \ell^s(S)$. If we happen to know that $\ell^u(S) + \ell^s(S) = \ell^u(M_2) + \ell^s(M_2)$ then we get $\ell(M_3) \geq \ell^u(S) - \ell^u(M_2)$ and therefore

$$\max\{\ell(M_1), \ell(M_3)\} \geq |\ell^u(S) - \ell^u(M_2)|.$$

Although this last assumption seems at least plausible in certain cases, as the above propositions 7.6 and 7.7 indicate, it seems difficult to verify without additional non-degeneracy assumptions on S (which we want to avoid). Also we would like to obtain a stronger conclusion, namely an estimate for the length of the Conley indices of M_1 and M_3. This would insure that the sets M_1 and M_3 are stable under perturbations of the flow. With our application to bifurcation theory in mind we are willing to assume that M_2 is "nice" (e.g. a neighborhood retract) and that the whole set S lies in a small neighborhood of M_2.

7.8 Theorem:

Let (M_1, M_2, M_3) be a G-Morse decomposition of the isolated invariant set S. Then the following holds.

a) $\ell\big(C(M_1)\big) \geq \ell^u(M_2) - \ell^u(S)$.

b) If S has an isolating neighborhood N such that the inclusion $M_2 \hookrightarrow N$ induces a monomorphism $h^*(N) \longrightarrow h^*(M_2)$ then $\ell\big(C(M_3)\big) \geq \ell^u(S) - \ell^u(M_2)$. Therefore we obtain in this case

$$\max\big\{\ell\big(C(M_1)\big), \ell\big(C(M_3)\big)\big\} \geq |\ell^u(S) - \ell^u(M_2)|.$$

Proof:

Let $N_0 \subset N_1 \subset N_2 \subset N_3$ be a G-invariant *index filtration* for the G-Morse decomposition (M_1, M_2, M_3). This means that for $1 \leq j \leq k \leq 3$ the pair (N_k, N_{j-1}) is a G-index pair for the isolated invariant set $\{x \in S: \alpha(x) \cup \omega(x) \subset \bigcup_{i=j}^k M_i\}$. The construction of index filtrations as in [Sa], Corollary 4.4, works as well in the equivariant world.

a) Since $C(M_1) = N_1/N_0$ we have to show $\ell(N_1, N_0) \geq \ell^u(M_2) - \ell^u(S)$. Because of Proposition 7.3 we may assume that $\ell^u(S) = \ell(N_0)$. Furthermore this proposition yields $\ell(N_1) \geq \ell^u(M_2)$ since (N_2, N_1) is a G-index pair for M_2. Applying the triangle inequality (see Theorem 4.7) we obtain therefore

$$\ell\big((C(M_1)\big) = \ell(N_1, N_0) \geq \ell(N_1) - \ell(N_0) \geq \ell^u(M_2) - \ell^u(S).$$

b) Since N is an isolating G-neighborhood of S there exists a G-invariant index filtration $N_0 \subset N_1 \subset N_2 \subset N_3$ as above with $N_3 = N$. Having chosen N_3 we cannot assume $\ell^u(S) = \ell(N_0)$. Set $k := \ell(C(M_3)) = \ell(N_3, N_2)$ and $m := \ell^u(M_2)$. We have to show $\ell^u(S) \leq k + m$. Choose elements $A_1, \ldots, A_k$ of $\mathcal{A}$ as in the definition of $\ell(N_3, N_2)$ and $A_{k+1}, \ldots, A_{k+m} \in \mathcal{A}$ as in the characterization of $\ell^u(M_2)$ in Theorem 7.5. For $i = 1, \ldots, k + m$ let $\omega_i \in \ker\big(R \longrightarrow h^*(A_i)\big)$ be given. According to 7.5 there exists $\alpha \in h^*(N_2, N_1) \cong h^*\big(C(M_2)\big)$ with $\alpha|M_2 = \omega_{k+1} \cdot \ldots \cdot \omega_{k+m} \cdot 1_{M_2}$. Here $|M_2$ denotes the homomorphism $h^*(N_2, N_1) \longrightarrow h^*(M_2)$ induced by the inclusion $M_2 \hookrightarrow N_2 \hookrightarrow (N_2, N_1)$. We shall use similar notation in the sequel. Applying the connecting homomorphism $\delta: h^*(N_2, N_1) \longrightarrow h^*(N_3, N_2)$ we get

$$\delta(\omega_1 \cdot \ldots \cdot \omega_k \cdot \alpha) = \omega_1 \cdot \ldots \cdot \omega_k \cdot \delta(\alpha) = 0 \in h^*(N_3, N_2)$$

by definition of $k = \ell(N_3, N_2)$. Therefore there exists $\beta \in h^*(N_3, N_1)$ with

$$\beta|(N_2, N_1) = \omega_1 \cdot \ldots \cdot \omega_k \cdot \alpha \in h^*(N_2, N_1).$$

We claim that $\beta\big|S = \omega \cdot 1_S \in h^*(S)$ where $\omega = \omega_1 \cdot \ldots \cdot \omega_{k+m} \in R$. Assuming this, Theorem 7.5 yields $\ell^u(S) \leq k + m$ as required because $\omega \cdot 1_S$ is contained in the image of $i^*\colon h^*(N_3, N_0) \longrightarrow h^*(S)\colon \omega \cdot 1_S = i^*\big(\beta\big|(N_3, N_0)\big)$. In order to prove the claim observe that $\beta\big|M_2 = \omega \cdot 1_{M_2} \in h^*(M_2)$ by our construction of α. The injectivity of the homomorphism $h^*(N_3) \longrightarrow h^*(M_2)$ implies $\beta\big|N_3 = \omega \cdot 1_{N_3}$ and therefore $\beta\big|S = \omega \cdot 1_{N_3}\big|S = \omega \cdot 1_S$. $\square$

7.9 Remark:

a) If $G = \mathbb{Z}/p$ then the theorem also holds for $\ell = \ell_0 + \ell_1 - 1$ as defined in Remark 4.14 provided $\ell_1^u(M_2) = \ell_0^u(M_2) + 1$. To prove this we use the notation of Theorem 7.8 and its proof. In part a) we have either $\ell_i(N_1) = \ell_i^u(M_2)$ for $i = 0, 1$ or $\ell_0(N_1) \geq \ell_0^u(M_2) + 1$. In the first case $\ell_1(N_1) = \ell_0(N_1) + 1$ so that the triangle inequality $\ell(N_1) \leq \ell(N_1, N_0) + \ell(N_0)$ holds (see Remark 4.14). And in the second case we have

$$\ell^u(M_2) \leq \ell(N_1) - 1 \leq \ell(N_1, N_0) + \ell(N_0).$$

In part b) we have either $\ell_1(N_3, N_2) = \ell_0(N_3, N_2) + 1$ or $\ell_1(N_3, N_2) = \ell_0(N_3, N_2)$. In the first case it follows from 7.8b) applied to ℓ_0 that

$$\ell^u(S) \leq 2\ell_0^u(S) \leq 2\big(\ell_0(N_3, N_2) + \ell_0^u(M_2)\big) = \ell(N_3, N_2) + \ell^u(M_2).$$

In the second case set $k := \ell_0(N_3, N_2)$ and $m := \ell_0^u(M_2)$. We have to show that $\ell^u(S) \leq 2k - 1 + 2m$, or equivalently $\ell_1^u(S) \leq k + m$ because $\ell_0^u(S) \leq \ell_0(N_3, N_2) + \ell_0^u(M_2) = k + m$. It suffices to prove that $c^{k+m-1} \cdot w \in h^*(\mathrm{pt}) = H_G^*(\mathrm{pt}; \mathbb{F}_p)$ lies in the image of $i^*\colon h^*(N_3, N_0) \longrightarrow h^*(S)$. This follows as in the proof of 7.8b) with $\omega_1 = w$ and $\omega_2 = \ldots = \omega_{k+m} = c$.

b) A very similar situation as the one of Theorem 7.8 has been considered by Floer and Zehnder in [FlZ]. There $G = S^1$ acts orthogonally on $X = \mathbb{R}^n$ with $X^G = 0$. The authors study an equivariant gradient flow on X which has $0 \in X$ as a hyperbolic fixed point with Morse index d. This fixed point corresponds to the set M_2. Therefore $\ell^u(M_2) = \ell^u(\{0\}) = d/2$. Furthermore, there exists a G-neighborhood N_1 of 0 which isolates an invariant set $S = \mathrm{inv}(N_1)$. We may assume that N_1 is a ball. Floer and Zehnder prove the existence of at least $|m - d/2|$ stationary G-orbits in S different from 0. The number m is defined to be half the minimal dimension of an element $\alpha \in H_G^*\big(\mathcal{C}(S)\big)$ that generates a free R-submodule; see [FlZ], Lemma 5 and its proof. At first sight the meaning of m is not clear. But from our point of view this number m has another interpretation: It is the exit-length $\ell^u(S)$ of S. Since the exit-length is a measure of the size of the exit set this interpretation is more geometric than the definition of m in [FlZ].

In order to see $m \leq \ell^u(S) =: k$ remember that $R = H_G^*(\mathrm{pt}) \cong \mathbb{Q}[c]$ with $c \in H_G^2(\mathrm{pt})$. Floer and Zehnder use real instead of rational coefficients but this makes no difference. Using the description of the exit-length in Theorem 7.5 there exists $\alpha \in H_G^{2k}\big(\mathcal{C}(S)\big)$ which restricts to $\alpha\big|S = c^k \cdot 1_S \in H_G^*(S)$. Now $c^k \cdot 1_S$ generates a free R-submodule of $H_G^*(S)$ because $S^G = \{0\} \neq \emptyset$. Therefore α generates a free R-submodule of $H_G^*(S)$ which implies $m \leq \frac{1}{2}\dim(\alpha) = k$.

To prove the inverse inequality consider an element

$$\alpha \in H_G^{2m}\big(\mathcal{C}(S)\big) \cong H_G^{2m}(N_1, N_0)$$

that generates a free R-submodule. Since $\ell(N_0) \leq \ell(X - 0) < \infty$ no element of $H_G^*(N_0)$ generates a free R-submodule. It follows that α cannot lie in the image of the connecting homomorphism $H_G^*(N_0) \longrightarrow H_G^{*+1}(N_1, N_0)$. Therefore $\alpha|N_1 \neq 0 \in H_G^{2m}(N_1)$. Since $H_G^{2m}(N_1) \cong H_G^{2m}(\text{pt}) \cong \mathbb{Q}$ is generated (as a vector space over $\mathbb{Q}$) by $c^m \cdot 1_{N_1}$ we see that we may assume $c^m \cdot 1_{N_1} = \alpha|N_1$. This implies $c^m \cdot 1_S = \alpha|S$, hence $\ell^u(S) \leq m$ as claimed.

7.5 A bifurcation theorem

We first study a finite-dimensional situation. Let $X = \mathbb{R}^n$ be a fixed point free orthogonal representation space of G and consider a family φ_λ of equivariant flows on X continuously parametrized by $\lambda \in \mathbb{R}$. We assume each φ_λ to be gradient-like with G-invariant Lyapunov-function $L_\lambda \colon X \to \mathbb{R}$. Since $0 \in X$ is the only point fixed by the action it must be a stationary solution of φ_λ for every $\lambda \in \mathbb{R}$. We are interested in bounded solutions of φ_λ which bifurcate from the trivial branch $\mathbb{R} \times \{0\}$. Fix some possible bifurcation point $\lambda_0 \in \mathbb{R}$ and assume that λ_0 is an isolated bifurcation point. Since φ is gradient-like this implies that there exists a neighborhood Λ of λ_0 in $\mathbb{R}$ such that φ_λ has the origin as an isolated stationary solution for every $\lambda \in \Lambda$ different from λ_0. Because of Theorem 7.4b) the number $\ell_-^u := \ell^u(\lambda, 0)$ for $\lambda < \lambda_0$ is independent of $\lambda \in \Lambda$. Similarly, the number $\ell_+^u := \ell_u(\lambda, 0)$ for $\lambda > \lambda_0$, $\lambda \in \Lambda$, is well defined. Here we use the notation from §7.3, that is $\ell^u(\lambda, 0) = \ell^u(0, \varphi_\lambda)$ is the exit-length of 0 with respect to the flow φ_λ.

7.10 Theorem:
  Suppose $\ell_-^u \neq \ell_+^u$. Then λ_0 is a bifurcation point, i.e. in every neighborhood of $(\lambda_0, 0)$ there exist stationary G-orbits (λ, Gx) with $x \neq 0$. Moreover, if the origin is also an isolated stationary solution for φ_{λ_0} (which means that all stationary solutions bifurcate either to the right or to the left of $\{\lambda_0\} \times X$) then there exists $\varepsilon > 0$ and integers $d_l, d_r \geq 0$ with $d_l + d_r \geq |\ell_+^u - \ell_-^u|$ so that for each $\lambda \in (\lambda_0 - \varepsilon, \lambda_0)$ respectively $\lambda \in (\lambda_0, \lambda_0 + \varepsilon)$ at least one of the following holds:

(i) *There exist stationary G-orbits Gx_λ^i, $i \in \mathbb{Z} - 0$, of φ_λ with $x_\lambda^i \neq 0$, $L_\lambda(x_\lambda^{-i}) < L_\lambda(0) < L_\lambda(x_\lambda^i)$ for $i \geq 1$, and $L_\lambda(x_\lambda^i) \to L_\lambda(0)$ as $|i| \to \infty$. In particular, φ_λ has infinitely many stationary G-orbits. They converge to $0 \in X$ as $\lambda \to \lambda_0$.*

(ii) *There exists an isolated invariant set $S_\lambda \subset X - 0$ with $\ell\big(\mathcal{C}(S_\lambda)\big) \geq d_l$ respectively $\ell\big(\mathcal{C}(S_\lambda)\big) \geq d_r$. Moreover, S_λ converges to $0 \in X$ as $\lambda \to \lambda_0$. i.e. for every neighborhood U of 0 in X there exists $\delta = \delta(U) > 0$ such that $S_\lambda \subset U$ if $|\lambda - \lambda_0| < \delta$.*

The result is also true for $G = \mathbb{Z}/p$ and $\ell = \ell_0 + \ell_1 - 1$ as in Remark 4.14.

Proof:

Suppose 0 is an isolated stationary solution of φ_{λ_0}. Choose an isolating G-neighborhood U of 0 in X which we may take to be a small ball. Then U is also isolating for φ_λ if λ is close to λ_0. Set $S'_\lambda := \mathrm{inv}(U, \varphi_\lambda)$, so that $S'_{\lambda_0} = \{0\}$. We define $d_l := |\ell^u(S'_{\lambda_0}) - \ell^u_-|$ and $d_r := |\ell^u(S'_{\lambda_0}) - \ell^u_+|$. Obviously, $d_l + d_r \geq |\ell^u_+ - \ell^u_-|$. According to Theorem 7.4b) there exists $\varepsilon > 0$ such that $\ell^u(S'_\lambda)$ is independent of λ for $|\lambda - \lambda_0| < \varepsilon$. Fix $\lambda \in (\lambda_0 - \varepsilon, \lambda_0)$ and define invariant subsets M_1, M_2 and M_3 of S'_λ as follows. The set M_1 consists of all stationary G-orbits Gx in S'_λ with $L(x) > L(0)$ and all connecting orbits between these. The set M_2 consists only of the origin. And the set M_3 contains all stationary G-orbits Gx in $S'_\lambda - 0$ with $L(x) \leq L(0)$, and all connecting orbits between these. If $\{M_1, M_2, M_3\}$ is not a G-Morse decomposition of S'_λ then there must exist a sequence Gx^i_λ, $i \geq 1$, of stationary G-orbits of φ_λ in U such that $L_\lambda(x^i_\lambda) > L_\lambda(0)$ and $L_\lambda(x^i_\lambda) \to L_\lambda(0)$ as $i \to \infty$. In that case we define new invariant subsets M'_1 and M'_3 of $S'_\lambda - 0$ as follows. The set M'_1 consists of all stationary G-orbits Gx in $S'_\lambda - 0$ with $L(x) \geq L(0)$ and all connecting orbits between these. And M'_3 consists of all stationary G-orbits Gx in S'_λ with $L(x) < L(0)$ and all connecting orbits between these. If $\{M'_1, M_2, M'_3\}$ is not a G-Morse decomposition of S'_λ then one obtains a sequence Gx^i_λ, $i \leq -1$, of stationary G-orbits of φ_λ in U satisfying $L_\lambda(x^i_\lambda) < L_\lambda(0)$ and $L_\lambda(x^i_\lambda) \to L_\lambda(0)$ as $i \to -\infty$. Thus (i) holds.

If (i) does not hold then the above considerations yield a G-Morse decomposition $\{M_1, M_2, M_3\}$ of S'_λ with $M_2 = \{0\}$. In that case we can apply Theorem 7.8 (respectively Remark 7.9a) and obtain

$$\max\left\{\ell\big(\mathcal{C}(M_1)\big), \ell\big(\mathcal{C}(M_3)\big)\right\} \geq |\ell^u(S'_\lambda) - \ell^u_-| = d_l.$$

Thus we may set $S_\lambda = M_1 \cup M_3$. We argue analogously for $\lambda \in (\lambda_0, \lambda_0 + \varepsilon)$ and d_r instead of d_l.

The convergence of S_λ towards 0 is clear because the neighborhood U chosen in the beginning can be arbitrarily small.   □

Due to the continuation invariance of the Conley index λ_0 must be a bifurcation point if the Conley index of 0 for φ_λ changes as λ passes λ_0. The exit-length allows to measure this change. Observe that $\ell\big(\mathcal{C}(0, \varphi_\lambda)\big) = \ell(\mathrm{pt})$ independently of λ because 0 is a fixed point of the group action.

Since $\ell(S_\lambda) \geq \ell\big(\mathcal{C}(S_\lambda)\big)$ we may apply the theorems 6.1 and 6.3 to obtain results on the number of stationary solutions and connecting orbits on S_λ. In addition we know that S_λ can be continued. We conjecture that in Theorem 7.10 there always exists a compact invariant set $S_\lambda \subset X - 0$ with $\ell(S_\lambda) \geq d_l$ for $\lambda \in (\lambda_0 - \varepsilon, \lambda_0)$ and $\ell(S_\lambda) \geq d_r$ for $\lambda \in (\lambda_0, \lambda_0 + \varepsilon)$ provided all stationary solutions bifurcate either to the right or to the left of $\{\lambda_0\} \times X$. These sets will in general not be isolated. We can prove this for $G = \mathbb{Z}/p$ or if $G = S^1 \times \Gamma$ with Γ finite.

7.11 Theorem:
Suppose $G = \mathbb{Z}/p$, p a prime, or $G = S^1 \times \Gamma$, Γ a finite group, and let ℓ be any of the lengths defined in 4.4, 4.14 or 4.16. Let φ be an equivariant product flow on $\mathbb{R} \times X$ where X is a fixed point free orthogonal representation space of G. Suppose that the origin $0 \in X$ is an isolated stationary solution of φ_λ also for $\lambda = \lambda_0$. Let ℓ^u_- and ℓ^u_+ be the exit-lengths as above. Then there exist $\varepsilon > 0$ and integers $d_l, d_r \geq 0$ with $d_l + d_r \geq |\ell^u_- - \ell^u_+|$ such that the following holds: For each $\lambda \in (\lambda_0 - \varepsilon, \lambda_0)$ respectively $\lambda \in (\lambda_0, \lambda_0 + \varepsilon)$ there exists a compact invariant set $S_\lambda \subset X - 0$ with $\ell(S_\lambda) \geq d_l$ respectively $\ell(S_\lambda) \geq d_r$.

Proof:
Choose a small open ball U around 0 in X which isolates 0 for the flow φ_{λ_0}. As in the proof of Theorem 7.10 there exists $\varepsilon > 0$ such that U is also isolating for $|\lambda - \lambda_0| < \varepsilon$ and $\ell^u(S'_\lambda)$, $S'_\lambda := \operatorname{inv}(U, \varphi_\lambda)$, is independent of λ. We define d_l and d_r as in the proof of Theorem 7.10. Fix $\lambda \in (\lambda_0 - \varepsilon, \lambda_0)$. We distinguish two cases depending on whether $\ell^u(S'_\lambda)$ is less or greater than ℓ^u_-. Assume first $\ell^u(S'_\lambda) \leq \ell^u_-$. Choose an index pair (M_1, M_0) of S'_λ with $M_1 \subset U$ and $\ell(M_0) = \ell^u(S'_\lambda)$. Next choose an index pair (N_1, N_0) of 0 for the flow φ_λ such that $N_1 \subset M_1 - M_0$ isolates 0 and $\ell(N_0) = \ell^u(0, \varphi_\lambda) = \ell^u_-$. Then we set

$$A := \{x \in N_0 : \omega(x) \subset S'_\lambda\}$$

and $S_\lambda := \omega(A) \subset S'_\lambda$. Now A and S_λ are closed G-subsets of U and S_λ is compact invariant. Each $x \in N_0 - A$ must eventually leave U (hence M_1) because otherwise it would limit in S'_λ. From this it follows that $\ell(N_0 - A) \leq \ell^u(S'_\lambda) = \ell^u(S'_{\lambda_0})$. Then using the continuity and the subadditivity property of the length

$$\ell^u_- = \ell(N_0) \leq \ell(A) + \ell(N_0 - A) \leq \ell(A) + \ell^u(S'_{\lambda_0}).$$

On the other hand, $\ell(S_\lambda) \geq \ell(A)$ since for any neighborhood V of S_λ there exists a time t such that $\varphi(A \times \{t\}) \subset V$. Thus we obtain $\ell(S_\lambda) \geq \ell^u_- - \ell^u(S'_{\lambda_0}) = d_l$.

If $\ell^u(S'_\lambda) \geq \ell^u_-$ then we replace φ by the inverse flow φ^-. Proposition 7.7 implies

$$\ell^u(S'_{\lambda_0}, \varphi^-) = \ell^s(S'_{\lambda_0}, \varphi) = \ell(SX) - \ell^u(S'_{\lambda_0}, \varphi)$$

and analogously $\ell^u_-(\varphi^-) = \ell^s_-(\varphi) = \ell(SX) - \ell^u_-(\varphi)$. Moreover, one can show (similarly to the proof of 7.7) that $\ell^s(S'_\lambda, \varphi_\lambda) + \ell^u(S'_\lambda, \varphi_\lambda) = \ell(SX)$ if λ is close to λ_0. Thus we obtain

$$\ell^u(S'_\lambda, \varphi^-_\lambda) = \ell^s(S'_\lambda, \varphi_\lambda) = \ell(SX) - \ell^u(S'_\lambda, \varphi_\lambda) \leq \ell(SX) - \ell^u_- = \ell^u_-(\varphi^-)$$

so that we can argue as above. This yields a compact invariant set S_λ with

$$\ell(S_\lambda) \geq \ell^u_-(\varphi^-) - \ell^u(S'_{\lambda_0}, \varphi^-_{\lambda_0}) = \ell^u(S'_{\lambda_0}, \varphi_{\lambda_0}) - \ell^u_-(\varphi) = d_l.$$

An analogous argument works for $\lambda \in (\lambda_0, \lambda_0 + \varepsilon)$. $\square$

Now we consider an infinite-dimensional situation that can be reduced to the Theorems 7.10 and 7.11. Let X be a G-Hilbert space and consider the differential equation

$$(*) \qquad\qquad \dot{x} = f_\lambda(x) = f(\lambda, x)$$

parametrized by $\lambda \in \mathbb{R}$. We assume the following.

(1) $f: \mathbb{R} \times X \longrightarrow X$ is the gradient (with respect to $x \in X$) of a G-invariant C^2-function $F: \mathbb{R} \times X \longrightarrow \mathbb{R}$, $f_\lambda(x) = \nabla F_\lambda(x)$. In particular, f_λ is equivariant.

(2) $f_\lambda(0) = 0$ for every $\lambda \in \mathbb{R}$.

If $X^G = 0$ then $f_\lambda(0) = 0$ is a consequence of the equivariance of f_λ. Observe that $A_\lambda := Df_\lambda(0): X \to X$ is equivariant because the origin is fixed by the group action. Hence, $\ker(A_\lambda)$ is a G-invariant linear subspace of X.

(3) For some $\lambda_0 \in \mathbb{R}$ the linearization A_{λ_0} has 0 as an eigenvalue with finite multiplicity, isolated in the spectrum $\sigma(A_{\lambda_0})$. Moreover, A_λ is an isomorphism if $\lambda \neq \lambda_0$ is close to λ_0.

(4) G acts without fixed points on $X_c := \ker(A_{\lambda_0})$, that is $X_c^G = 0$.

Let $\sigma_0(A_\lambda)$ denote the 0-group of A_λ. It consists of all eigenvalues of A_λ which approach 0 as $\lambda \to \lambda_0$; $\sigma_0(A_\lambda)$ is defined for λ near λ_0 (see [Ka], II.2.1 and IV.3.5). Define X_λ to be the generalized eigenspace of A_λ belonging to $\sigma_0(A_\lambda) \cap \mathbb{R}^-$. This is again an invariant subspace of X. The spaces X_λ and X_μ are G-isomorphic if $\lambda - \lambda_0$ and $\mu - \lambda_0$ have the same sign. In particular, $\ell(SX_{\lambda_0^-}) := \ell(SX_{\lambda_0 - \delta})$ and $\ell(SX_{\lambda_0^+}) := \ell(SX_{\lambda_0 + \delta})$ are independent of $\delta > 0$ small. Let φ_λ denote the flow on X associated to $(*)$.

7.12 Theorem:

If (1) – (4) hold then either 0 is not an isolated stationary orbit of φ_{λ_0} or there exist natural numbers d_l and d_r with $d_l + d_r \geq |\ell(SX_{\lambda_0^-}) - \ell(SX_{\lambda_0^+})|$ and satisfying the alternative "(i) or (ii)" of Theorem 7.10 (with F instead of L). Moreover, if $G = \mathbb{Z}/p$ or $G = S^1 \times \Gamma$, Γ a finite group, then the conclusion of Theorem 7.11 holds. In these cases we work with the lengths defined in 4.4, 4.14 or 4.16.

Combining this result with the computations of the length of representation spheres in §5.2 and Theorem 6.3 respectively Remark 6.4a) we obtain our next result.

7.13 Corollary:
Suppose (1)–(4) hold and 0 is an isolated stationary orbit of φ_{λ_0}.

a) *If $G = \mathbb{Z}/p$, p a prime, then there exist natural numbers d_l and d_r with*

$$d_l + d_r \geq |\dim X_{\lambda_0^-} - \dim X_{\lambda_0^+}|$$

and the following property. For $\lambda < \lambda_0$ close to λ_0 the flow φ_λ has at least d_l stationary G-orbits different from 0. They converge to 0 if $\lambda \to \lambda_0$. If there are only finitely many stationary G-orbits of φ_λ close to 0 then there exist at least d_l stationary G-orbits Gx_i, $i = 1, \ldots, d_l$, and connecting orbits from Gx_{i+1} to Gx_i. The bifurcating set consisting of stationary G-orbits and connecting orbits between them has dimension at least $d_l - 1$. Similarly for $\lambda > \lambda_0$ and d_r instead of d_l.

b) *If $G = S^1 \times \Gamma$, Γ finite, the same is true with $d_l + d_r \geq \frac{1}{2}|\dim X_{\lambda_0^-} - \dim X_{\lambda_0^+}|$. Here the dimension of the bifurcating set is at least $2d_l - 1$ respectively $2d_r - 1$.*

Proof of Theorem 7.12:
We deduce 7.12 from 7.10 and 7.11 using a finite-dimensional reduction. Because we are not only interested in the stationary G-orbits of φ_λ the Lyapunov-Schmidt reduction does not suffice. Instead we apply the *center manifold theorem*. The existence of a G-invariant center manifold in the situation of Theorem 7.12 has been proved in a paper by Vanderbauwhede and Iooss [VI]. Here we obtain the following: Let $\pi_c\colon X \to X_c$ be the orthogonal projection and set $X_h := \ker(\pi_c) = X_c^\perp$. There exists a G-neighborhood Ω of the origin in X, a neighborhood Λ of λ_0 in $\mathbb{R}$ and an equivariant map $\alpha\colon \Lambda \times X_c \longrightarrow X_h$ with $\alpha_\lambda := \alpha(-, \lambda)$ differentiable, $\alpha_\lambda(0) = 0$ for all λ and $D\alpha_{\lambda_0} = 0$. These data satisfy the following:

— If $x\colon I \to X_c$ is a solution of

(∗∗) $$\dot{x} = \pi_c \circ f_\lambda\big(x + \alpha_\lambda(x)\big)$$

and $\tilde{x}(t) := x(t) + \alpha_\lambda\big(x(t)\big) \in \Omega$ for all $t \in I$ then $\tilde{x}$ is a solution of (∗).

— If $\tilde{x}\colon \mathbb{R} \to X$ is a solution of (∗) such that $\tilde{x}(t) \in \Omega$ for all $t \in \mathbb{R}$ then

$$\tilde{x}(t) - \pi_c \circ \tilde{x}(t) = \alpha_\lambda\big(\pi_c \circ \tilde{x}(t)\big) \text{ for all } t \in \mathbb{R}$$

and $x := \pi_c \circ \tilde{x}\colon \mathbb{R} \to X_c$ is a solution of (∗∗).

$M_c := \text{Graph}(\alpha) \cap (\Lambda \times \Omega)$ is a (local) center manifold associated to (∗). Let ψ be the (local) equivariant flow on X_c induced by (∗∗). The projection $\pi_c\colon M_c \to X_c$ provides an equivariant flow equivalence between $\varphi|M_c$ and ψ. Obviously ψ is gradient-like since this is true for φ. (It is not clear whether ψ is a gradient map if φ is one.) Theorem 7.12 is now a consequence of Theorems 7.10 and 7.11 applied to ψ. One simply observes that for $\lambda \neq \lambda_0$ the tangent space of the unstable manifold of the isolated stationary orbit 0 of ψ_λ is just $\pi_c(X_\lambda)$. Since $\pi_c|X_\lambda$ is an isomorphism we obtain

$$\ell^u(0, \psi_\lambda) = \ell\big(S(\pi_c(X_\lambda))\big) = \ell(SX_\lambda).$$

□

Let us comment a bit on these results and their proofs.

7.14 Remark:

a) Even if one is only interested in the bifurcating stationary solutions 7.10 – 7.13 generalize and improve older results. As mentioned above, for $G = \mathbb{Z}/2$ Fadell and Rabinowitz [FaR1] studied bifurcation of critical points of an even map $F_\lambda \colon X \to \mathbb{R}$ under the additional assumption $\nabla F_\lambda(x) = Ax - \lambda x + o(\|x\|)$. If λ_0 is an isolated eigenvalue of A with multiplicity m then at least m pairs of critical points of F bifurcate from $(\lambda_0, 0)$. Similarly, in [FaR2] the case $G = S^1$ has been studied and applied to prove the existence of periodic solutions of Hamiltonian systems near a stationary solution. We shall generalize this in chapter 9. In both papers Fadell and Rabinowitz apply minimax methods using the length (the cohomological index in their terms) for $G = \mathbb{Z}/2$ and S^1. In [BC1] Bartsch and Clapp use a generalized mountain pass theorem to obtain a lower bound for the number of bifurcating critical G-orbits. Already for $G = \mathbb{Z}/p$, p an odd prime, we improve the result of [BC1] by the factor 2. Our approach is closer in spirit to the one of Floer and Zehnder [FlZ]. They consider the case $G = S^1$ in order to give a new proof of the result of [FaR2]. They also use the equivariant Conley index and Borel cohomology but not the S^1-length. Nevertheless, the exit-length appears implicitly in Lemma 5 of [FlZ]; see also Remark 7.9b).

b) The main additional ingredient in the proof of the infinite-dimensional results 7.12 and 7.13 is the center manifold theorem. A G-invariant center manifold exists in much more general situations than those considered here. For example, look at the equation

$$(*)\qquad\qquad \dot{x} = f_\lambda(x) = A_\lambda \cdot x + o(\|x\|).$$

Here X is a G-Banach space and $A_\lambda \colon D(A_\lambda) \subset X \to X$ is a linear operator. A solution of $(*)$ for a given parameter λ is a continuous map $x \colon I \to D(A_\lambda)$, where I is an open interval, satisfying:

— The composition $I \xrightarrow{\ x\ } D(A_\lambda) \hookrightarrow X$ is differentiable.

— $\dot{x}(t) = f_\lambda\big(x(t)\big)$ for every $t \in I$.

If $A_\lambda \colon D(A_\lambda) \subset X \to X$ is a densely defined closed linear operator a center manifold exists provided the following hypotheses are satisfied:

— $\sigma(A_{\lambda_0}) \cap i\mathbb{R}$ consists of a finite number of isolated eigenvalues, each with a finite-dimensional eigenspace. Moreover, $\sigma(A_\lambda) \cap i\mathbb{R} = \emptyset$ for every $\lambda \neq \lambda_0$.

— There exist constants $a \in \mathbb{R}$, $\delta > 0$ and $c > 0$ such that for all $\mu \in \mathbb{C}$:

$$\mathrm{Re}(\mu) \le a - \delta|\mathrm{Im}(\mu)| \quad \Rightarrow \quad \mu \in \mathbb{C} - \sigma(A_{\lambda_0}) \text{ and } \|\mu\mathrm{Id} - A_{\lambda_0}\|^{-1} \le c/(1 + |\mu|).$$

This covers the case where A_λ generates an analytic semigroup (see [Henry] or [Pazy]). The existence of a G-invariant center manifold under much more general hypotheses has been proved in [VI]. It is not necessary that A_λ generates an analytic semigroup or that the Cauchy problem associated to $(*)$ is well posed or that backward solutions exist etc.

c) We can also study a much more degenerate situation than the one considered here (see [KMP] in the non-equivariant setting). We need only assume that the origin is an isolated stationary solution of φ_{λ_i} and φ_{μ_i} where $\lambda_i < \lambda_0 < \mu_i$ are sequences convergent towards λ_0. Then a similar result holds provided $\ell^u(\lambda_i, 0) \neq \ell^u(\mu_i, 0)$. Here λ_0 need not be an isolated bifurcation point and even the λ_i and μ_i may be bifurcation points.

d) Using the continuation results of §6.4 one can proof a global version of Theorem 7.10. If part (ii) applies then either the bifurcating sets can be connected to a compact invariant set C_λ containing the origin for some $\lambda \neq \lambda_0$ or their quasi-component in the space of isolated invariant sets is not contained in a bounded subset of $\mathbb{R} \times X$. This latter statement does not imply that the bifurcating stationary solutions are unbounded. It may very well be that all bifurcating stationary solutions are contained in a small neighborhood of $(\lambda_0, 0)$ but that a family of connecting orbits $\left(\lambda_i, \varphi_{\lambda_i}(x_i, \mathbb{R})\right) \subset C_{\lambda_i}$ exists with $\bigcup_i \varphi_{\lambda_i}(x_i, \mathbb{R})$ not contained in a bounded subset.

Theorems 7.10 and 7.11 can also be considered as generalizations of the "invariant sphere theorem" of Field; see [Field], Theorem 5.1. He considers the following situation. Let $Q : \mathbb{R}^n \to \mathbb{R}^n$ be a homogeneous polynomial of degree $2d + 1$ $(d > 0)$ and assume that $\langle Q(x), x \rangle < 0$ for every $x \in \mathbb{R}^n - 0$. Consider the flow φ_λ on $\mathbb{R}^n$ of the differential equation

$$\dot{x} = f_\lambda(x) := \lambda x + Q(x).$$

Observe that f_λ and φ_λ are odd functions. The *invariant sphere theorem* now says that for every $\lambda > 0$ there exists a unique $(n-1)$-dimensional sphere $S_\lambda \subset \mathbb{R}^n - 0$ which is invariant by φ_λ. Field also shows that these invariant spheres continue to exist under certain perturbations of the flow ([Field], Theorem 5.23). More precisely, let ψ_λ be the flow associated to

$$(+) \qquad\qquad \dot{x} = \lambda x + Q(x) + P_\lambda(x)$$

where either $P_\lambda(x) = O(\|x\|^{2d+2})$ for all λ or $P_\lambda(x) = O(\|x\|^2)$ and $P_0 \equiv 0$. In addition, Field assumes certain nondegeneracy assumptions on the vector field induced by Q on the sphere S^{n-1} (Axiom A, normal hyperbolicity of the bifurcating sets); essentially he requires this vector field to be structurally stable. It is easy to check that ψ_λ satisfies the following conditions:

(a) The origin is an isolated stationary orbit of ψ_λ for every λ near 0.

(b) For $\lambda \leq 0$ the origin is an attractor and for $\lambda > 0$ it is a repeller.

(c) All bifurcating branches are supercritical, i.e. there exists $\varepsilon > 0$ such that for $\lambda \in [-\varepsilon, 0]$ the origin is the only bounded solution in $\{x \in \mathbb{R}^n : \|x\| \leq \varepsilon\}$.

As an immediate consequence of Theorem 7.11 and its proof we obtain the following result.

7.15 Theorem:

Let $G = \mathbb{Z}/p$ act orthogonally on $X = \mathbb{R}^n$ and without fixed points except the origin. Consider an equivariant locally Lipschitz map $f_\lambda: X \to X$ such that the flow φ_λ associated to the differential equation $\dot{x} = f_\lambda(x)$ satisfies (a), (b) and (c). Then for every neighborhood U of the origin in $\mathbb{R}^n$ there exists $\delta > 0$ such that the following holds: For every $\lambda \in (0, \delta]$ there exists an isolated invariant set $S_\lambda \subset U - 0$ with $\ell(S_\lambda) = n = \ell\big(\mathcal{C}(S_\lambda)\big)$; here ℓ is as in 4.4 or 4.14. In particular, S_λ separates X in (at least) two path components. The origin is in a bounded component of $\mathbb{R}^n - S_\lambda$. Moreover, all trajectories of φ_λ through $x \in U - 0$ are forward asymptotic to S_λ.

Of course, without additional assumptions we cannot conclude that S_λ is homeomorphic to a sphere. But it is possible to show that the invariant sets continue to exist if we perturb f_λ by a function P_λ as in (+) which is not odd. In that case neither S_λ nor $\mathcal{C}(S_\lambda)$ are G-spaces. In Remark 6.4c) we indicated how one can use the knowledge in the equivariant setting to obtain results which are invariant under non-equivariant perturbations. The theorems 7.10 and 7.11 treat situations where neither (b) nor (c) need be satisfied. Moreover, the flow can be very degenerate and the oddness assumption is replaced by more general symmetries.

Chapter 8

Bifurcation for O(3)-equivariant problems

8.1 Introduction

In this chapter we study the bifurcation of stationary solutions and connecting orbits for $O(3)$-equivariant problems which are gradient-like. The symmetry group $O(3)$ has received much attention in recent years because it appears naturally in various applications, for instance the buckling of a spherical shell (see [KnS]) or the convection of a fluid confined between two concentric spherical shells (see [Bu]). We refer the reader to the recent article [CLM] of Chossat, Lauterbach and Melbourne for further references to the literature as well as a survey of results on the bifurcation of stationary solutions.

The physical model leads to a semiflow on some Sobolev space depending on a distinguished parameter (the pressure in the buckling problem or the temperature in the fluids problem). For each parameter value there is a trivial solution fixed by the whole group $O(3)$ which loses stability as the parameter changes. We assume that we have performed a center manifold reduction in the neighborhood of a possible bifurcation point. This reduces the problem to the study of a family of equivariant flows φ_λ on a finite-dimensional representation space V of $O(3)$, continuously parametrized by $\lambda \in \mathbb{R}$. The origin of V is a stationary orbit of φ_λ for every $\lambda \in \mathbb{R}$. We also assume that φ_λ is gradient-like, i.e. there exists an $O(3)$-invariant Lyapunov function $V \to \mathbb{R}$ which decreases strictly along non-constant trajectories. Typically the action of $O(3)$ on V is irreducible which implies that $\dim V = 2l + 1$ is odd. Furthermore we suppose that $-1 \in O(3)$ acts on V as minus the identity. This is a natural assumption if l is odd (see [CLM]).

We know from the results of chapter 3 that $O(3)$ is not a "nice" group as far as the equivariant category or genus are concerned. Our approach consists of applying the length for the subgroup $\mathbb{Z}/2 \cong \{\pm 1\}$ of $O(3)$. The symmetry of the flow with respect to the full group is used in two ways: First, the flow must leave the fixed point subspaces V^H invariant, where H is any (closed) subgroup of $O(3)$. Second, the bifurcating sets of stationary solutions and connecting orbits are not only $\mathbb{Z}/2$-invariant but $O(3)$-invariant. It turns out that this method yields non-trivial results.

In §8.2 we collect a number of known results on the subgroups and the irreducible representations of $O(3)$ which are needed in the sequel. Then in §8.3 we study the bifurcation of stationary solutions. In particular we are interested in the number of bifurcating orbits with a given isotropy group. Applying our bifurcation theorem from the last section we shall finally prove the existence of various heteroclinic orbits connecting the bifurcating stationary orbits. Section 8.4 is devoted to this topic. So far, connecting orbits have only been studied for $l \leq 3$ (that is $\dim V_l \leq 7$); see [FiM], [Lau]. We recover the connecting orbits for $l = 3$ (the case $l = 1$ is trivial) and prove the existence of many more connecting orbits for odd

$l \geq 5$. We also compare our approach with the work of Fiedler and Mischaikow [FiM] and Lauterbach [Lau].

8.2 Subgroups and representations of $O(3)$

In order to make this chapter readable we need to present various data about the group $O(3)$ of orthogonal 3×3-matrices. The reader can find this material (and much more) in [CLM]. We follow the notation of this paper, in particular we write $\mathbb{Z}_m$ instead of $\mathbb{Z}/m$ for the cyclic group of m elements and D_m for the dihedral group of $2m$ elements.

The center of $O(3)$ is $\mathbb{Z}_2^c = \{\pm 1\}$ and $O(3)$ is isomorphic to $SO(3) \times \mathbb{Z}_2^c$. The (closed) subgroups of $O(3)$ are well known. There are three types of subgroups:

— subgroups of $SO(3)$;
— subgroups that contain -1;
— subgroups that neither are contained in $SO(3)$ nor contain -1.

The subgroups of $SO(3)$ are up to conjugacy $O(2)$, $SO(2)$, D_m ($m \geq 2$), $\mathbb{Z}_m$ ($m \geq 1$) and the symmetry groups I of the icosahedron, O of the octahedron and T of the tetrahedron (cf. Example 3.20). We usually think of $O(2)$ as being generated by the rotations $\theta \in [0, 2\pi)$ and κ where

$$\theta = \begin{pmatrix} \cos\theta & -\sin\theta & 0 \\ \sin\theta & \cos\theta & 0 \\ 0 & 0 & 1 \end{pmatrix} \quad \text{and} \quad \kappa = \begin{pmatrix} 1 & 0 & 0 \\ 0 & -1 & 0 \\ 0 & 0 & -1 \end{pmatrix}.$$

All other copies of $O(2)$ in $SO(3)$ are obtained via conjugation. Similar remarks hold for the other subgroups. Observe that we excluded the group $D_1 = \{1, \kappa\}$ from our list because it is conjugate to $\mathbb{Z}_2 = \{1, \pi\}$; one can use the matrix $\begin{pmatrix} 0 & 0 & 1 \\ 0 & 1 & 0 \\ 1 & 0 & 0 \end{pmatrix}$ for conjugation. We list these subgroups, their normalizes and their Weyl groups in Table 8.1.

8.1 Table:

Subgroups of $SO(3)$, their normalizes and Weyl groups in $O(3)$

H	NH	$WH = NH/H$
$O(2)$	$O(2) \times \mathbb{Z}_2^c$	$\mathbb{Z}_2$
$SO(2)$	$O(2) \times \mathbb{Z}_2^c$	$\mathbb{Z}_2 \times \mathbb{Z}_2$
$D_m,\ m \geq 3$	$D_{2m} \times \mathbb{Z}_2^c$	$\mathbb{Z}_2 \times \mathbb{Z}_2$
D_2	$O \times \mathbb{Z}_2^c$	$D_3 \times \mathbb{Z}_2$
$\mathbb{Z}_m,\ m \geq 2$	$O(2) \times \mathbb{Z}_2^c$	$O(2) \times \mathbb{Z}_2$
I	$I \times \mathbb{Z}_2^c$	$\mathbb{Z}_2$
O	$O \times \mathbb{Z}_2^c$	$\mathbb{Z}_2$
T	$O \times \mathbb{Z}_2^c$	$\mathbb{Z}_2 \times \mathbb{Z}_2$

If a subgroup H of $O(3)$ contains -1 then it is of the form $H = K \times \mathbb{Z}_2^c$ where K is a subgroup of $SO(3)$. In that case $NH = NK$ and WH may be trivial. Finally, all subgroups H of $O(3)$ with $-1 \notin H$ and $H \not\subset SO(3)$ are obtained in the following way. Let K be the image of H under the projection $\pi : O(3) \to SO(3)$ given by $\pi((-1) \cdot g) = g$ for every $g \in SO(3)$. The intersection $L := H \cap SO(3)$ is a subgroup of K of index 2 and $H = L \cup (-1) \cdot (K - L)$. Observe that H is isomorphic to K but not conjugate to it in $O(3)$. In Table 8.2 we list the subgroups of this type, the associated subgroups K and L, the normalizers and Weyl groups.

8.2 Table:
Subgroups of $O(3)$ not contained in $SO(3)$ nor containing -1

H	$K = \pi(H)$	$L = H \cap SO(3)$	NH	WH
$O(2)^-$	$O(2)$	$SO(2)$	$O(2) \times \mathbb{Z}_2^c$	$\mathbb{Z}_2$
D_{2m}^d, $m \geq 2$	D_{2m}	D_m	$D_{2m} \times \mathbb{Z}_2^c$	$\mathbb{Z}_2$
D_m^z, $m \geq 2$	D_m	$\mathbb{Z}_m$	$D_{2m} \times \mathbb{Z}_2^c$	$\mathbb{Z}_2 \times \mathbb{Z}_2$
$\mathbb{Z}_{2m}^-$	$\mathbb{Z}_{2m}$	$\mathbb{Z}_m$	$O(2) \times \mathbb{Z}_2^c$	$O(2)$
O^-	O	T	$O \times \mathbb{Z}_2^c$	$\mathbb{Z}_2$

The group $D_1^z = \{1, (-1) \cdot \kappa\}$ is conjugate to $\mathbb{Z}_2^- = \{1, (-1) \cdot \pi\}$ and the group $D_2^d = \{1, \kappa, (-1) \cdot \pi, (-1) \cdot \kappa \cdot \pi\}$ is conjugate to $D_2^z = \{1, \pi, (-1) \cdot \kappa, (-1) \cdot \kappa \cdot \pi\}$, so they are excluded from the table. Every subgroup of $O(3)$ which does not contain -1 is conjugate to precisely one group listed in Tables 8.1 and 8.2.

Now we study the representations of $O(3)$. Since $O(3) = SO(3) \times \mathbb{Z}_2^c$ the irreducible representations of $O(3)$ are precisely the irreducible representations of $SO(3)$ tensored with the irreducible representations of $\mathbb{Z}_2^c$. Thus each irreducible representation of $SO(3)$ gives rise to two irreducible representations of $O(3)$ depending on whether $-1 \in \mathbb{Z}_2^c$ acts trivially or not. As already stated in Example 3.20, the irreducible representations of $SO(3)$ are the spaces V_l, $l = 0, 1, 2, \ldots$, of spherical harmonics of order l. There is a natural action of $O(3)$ on V_l so that -1 acts trivially if l is even and is minus the identity on V_l if l is odd. (If $f = f(x) \in V_l$ then $(gf)(x) = f(g^{-1}x)$ for any $g \in O(3)$, $x \in \mathbb{R}^3$. And $f(-x) = (-1)^l f(x)$ because f is homogeneous of degree l.) It is this natural representation of $O(3)$ which appears in many applications. In the sequel we shall only consider odd l and the natural representation of $O(3)$ on V_l. The methods apply to all representations of $O(3)$ on which -1 acts as minus the identity; the representations need not be irreducible.

The final piece of information which we need is the lattice of conjugacy classes (H) of isotropy groups $H \subset O(3)$ of $V_l - 0$ and the dimensions of the fixed point subspaces V_l^H. The set of conjugacy classes of subgroups of $O(3)$ is partially ordered by subconjugation, that is $(H) \leq (K)$ iff $gHg^{-1} \subset K$ for some $g \in O(3)$. Since V_l is finite-dimensional there exist only finitely many conjugacy classes of isotropy subgroups. For $l = 3$ respectively $l = 5$ this lattice can be represented by the graphs of Figure 8.3 respectively 8.4.

8.3 Figure:
 The lattice of isotropy subgroups for $l = 3$. The number in the column on the right is the dimension of the corresponding fixed point spaces.

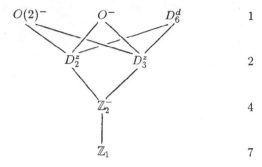

8.4 Figure:
 The lattice of isotropy subgroups for $l = 5$. The number in the column on the right is the dimension of the corresponding fixed point spaces.

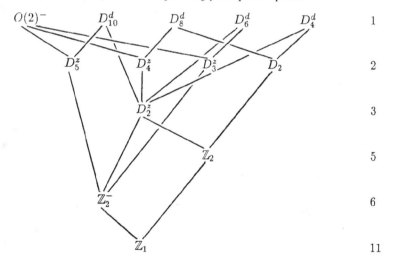

These graphs have to be interpreted in the following way. Look at Figure 8.3, for instance. For $H = O(2)^-$, O^- or D_6^d we have dim $V_3^H = 1$, for $H = D_2^z$ or D_3^z we have dim $V_3^H = 2$ and so on. Subgroups of $O(3)$ which are not conjugate to any vertex do not appear as isotropy group in $V_3 - 0$. If two vertices are connected by an edge then the lower subgroup is subconjugate to the upper one. For instance, $H = D_3^z$ is subconjugate to $K = O^-$, hence the two-dimensional fixed point space V_3^H contains the one-dimensional fixed point space $V_3^{gKg^{-1}} = g \cdot V_3^K$ if $H \subset gKg^{-1}$. In fact, in this special case there are two different subgroups $g_1 K g_1^{-1}$ and $g_2 K g_2^{-1}$ containing H. Thus V_3^H contains the two different fixed point subspaces $g_1 V_3^K$ and $g_2 V_3^K$. For some considerations the number of different conjugate copies of a subgroup K containing a subgroup H is important. We shall not list these numbers here, they can also be found in [CLM], Theorem A.2.

For $l > 5$ we only list the dimensions of the fixed point spaces in Table 8.5; $[x]$ denotes the greatest integer less than or equal to x.

8.5 Table:
 Isotropy subgroups of $V_l - 0$ and the dimensions of the fixed point spaces for odd $l > 5$.

H	dim V_l^H	condition on H and l
$O(2)^-$	1	
D_{2m}^d	$\left[\frac{l+m}{2m}\right]$	$2 \leq m \leq l$
D_m^z	$\left[\frac{l}{m}\right] + 1$	$2 \leq m \leq l$
D_m	$\left[\frac{l}{m}\right]$	$2 \leq m \leq l$
$\mathbb{Z}_{2m}^-$	$2\left[\frac{l+m}{m}\right]$	$1 \leq m \leq l/3$
$\mathbb{Z}_m$	$2\left[\frac{l}{m}\right] + 1$	$1 \leq m < l/2$
I	$\left[\frac{l}{5}\right] + \left[\frac{l}{3}\right] + \left[\frac{l}{2}\right] - l + 1$	$l = 15, 21, 25, 27, \ l \geq 31$
O^-	$\left[\frac{l}{3}\right] - \left[\frac{l}{4}\right]$	
O	$\left[\frac{l}{4}\right] + \left[\frac{l}{3}\right] + \left[\frac{l}{2}\right] - l + 1$	$l = 9, \ l \geq 13$
T	$2\left[\frac{l}{3}\right] + \left[\frac{l}{2}\right] - l + 1$	$l = 9, \ l \geq 13$

The relations $(H) \leq (K)$ for the isotropy subgroups in Table 8.5 can be found again in [CLM], Table A.5.

8.3 Bifurcating stationary solutions

We fix a finite-dimensional representation V of $G = O(3)$ on which -1 acts as minus the identity. Let φ_λ be a family of equivariant gradient-like flows on V continuously parametrized by $\lambda \in \mathbb{R}$. Since $0 \in V$ is the only point fixed by the action it must be a stationary point of φ_λ for every λ. We fix a possible bifurcation parameter λ_0 and assume that the origin is an isolated stationary solution of φ_λ for $\lambda \neq \lambda_0$ close to λ_0. For simplicity we even assume:

(H1) The origin is an attractor for $\lambda < \lambda_0$ and a repeller for $\lambda > \lambda_0$.

This hypothesis is reasonable when one thinks of the flow φ_λ as being obtained by a center-manifold reduction. In that case V is the kernel of an $O(3)$-equivariant linear map (between $O(3)$-Banach spaces), so V is generically irreducible. But then the only linear maps $V \to V$ which commute with the action of G are multiples of the identity, hence we may generically assume that $D\varphi_\lambda(0)x = (\lambda - \lambda_0)x$. We need one more hypothesis which also holds generically.

(H2) The origin is an isolated stationary point of φ_{λ_0}.

8.6 Theorem:

   Suppose (H1) and (H2) hold. Let $H \subset O(3)$ be a maximal isotropy group in $V - 0$. Then there exists $\epsilon > 0$ and integers $d_l(H), d_r(H) \geq 0$ with $d_l(H) + d_r(H) \geq \dim V^H$ and the following property. For each $\lambda \in (\lambda_0 - \epsilon, \lambda_0)$ resp. $\lambda \in (\lambda_0, \lambda_0 + \epsilon)$ the flow φ_λ has at least d_l resp. d_r stationary G-orbits homeomorphic to G/H. If $V = V_l$ is irreducible and l is odd this applies to $H = O(2)^-, O^-, O, I$ and D_{2m}^d with $l/3 < m \leq l$. The dimension of V_l^H can be found in Table 8.5.

Proof:

   The only possible maximal isotropy groups are $O(2)^-, O^-, O, I$ and $D_{2m}^d, m \geq 2$. For each of these groups WH is isomorphic to $\mathbb{Z}_2$ and acts freely on V^H. The fixed point spaces V^H are invariant under the flow φ_λ and $\varphi_\lambda|V^H$ is WH-equivariant. Therefore the claim follows from the theorems 7.10 and 6.3. If $V = V_l$ then according to Table 8.5 the dihedral groups D_{2m}^d with $2 \leq m \leq l$ appear as isotropy groups (except for $m = 2$ if $l = 3$). But only for $l/3 < m \leq l$ these groups are maximal because $D_{2m}^d \subset D_{2n}^d$ if n/m is an odd integer.   □

   If V is a fixed point free representation of G (which is the case of most interest to us) then the isotropy groups of the bifurcating solutions must be smaller than G. Speaking sloppily, the bifurcating solutions have less symmetry than the trivial one. This phenomenon is called *spontaneous symmetry breaking* as opposed to the *forced symmetry breaking*. The latter one occurs when an equation like $\dot{x} = f(x)$, or simply $f(x) = 0$, is perturbed by a term $\epsilon \cdot g(x)$; here f is supposed to be equivariant with respect to a group G whereas g is only equivariant with respect to a subgroup H of G. Then the solutions of the perturbed equation loose symmetry because already the equation itself does so.

   The question whether solutions with submaximal isotropy group exist is much more subtle. Consider for instance the tetrahedral group T which appears as isotropy group if $l = 9$ or $l \geq 13$ (see Table 8.5). Since $WT \cong \mathbb{Z}_2 \times \mathbb{Z}_2$ we can apply the results of chapters 6 and 7 and obtain at least $d(V, T) := \dim V^T$ stationary G-orbits with isotropy at least T. However, T is contained in O, O^- and I, and $d(V_l, T) \leq d(V_l, O) + d(V_l, O^-) + d(V_l, I)$ for every odd l. So we cannot predict more solutions with isotropy at least T than those homeomorphic to G/O, G/O^- or G/I. The same is true for the other submaximal isotropy groups except for the dihedral groups D_{2m}^d. As an example, look at the case $l = 9$ and D_6^d. We have $d(V_9, D_6^d) = 2$ but D_6^d is only subconjugate to one other isotropy group, namely D_{18}^d. Since $d(V_9, D_{18}^d) = 1$ we obtain one more solution with isotropy at least D_6^d than the one with isotropy D_{18}^d guaranteed by Theorem 8.6. In a degenerate situation we cannot exclude the case that $d_l(D_6^d) = d_r(D_6^d) = 1$ (in the notation of 8.6) and the bifurcating stationary G-orbits have both the maximal isotropy D_{18}^d. This is not possible if $d_l(D_6^d) = 2$ or $d_r(D_6^d) = 2$. We can say more in a certain special situation which occurs frequently in applications.

8.7 Theorem:

Let l be an odd integer and consider a G-invariant C^2-map $F: V_l \to \mathbb{R}$ with $\nabla F(0) = 0$. Let φ_λ be the flow associated to the ordinary differential equation $\dot{x} = \lambda x - \nabla F(x)$. Suppose λ_0 is an eigenvalue of the Hessian of F at 0 and that φ_λ satisfies (H1). Then for any $R > 0$ and any maximal isotropy group H of $V_l - 0$ there exist at least $\dim V_l^H$ stationary G-orbits Gx of $\varphi_{\lambda(x)}$ with $\|x\| = R$ and $G_x = H$. In addition, for each $l/9 < m \leq l/3$ there exists a stationary G-orbit Gx of $\varphi_{\lambda(x)}$ with $\|x\| = R$ and submaximal isotropy $G_x = D_{2m}^d$. Moreover, $\lambda(x) \to \lambda_0$ as $R \to 0$ for every such stationary G-orbit Gx; thus they bifurcate from $(\lambda_0, 0)$.

Proof:

We obtain the required stationary solutions as critical points of F constrained to $S_R V_l = \{x \in V_l : \|x\| = R\}$. Since $WH \cong \mathbb{Z}_2$ acts freely on $S_R V_l^H$ for the isotropy groups H in question we obtain at least $\{\mathbb{Z}_2\}\text{-cat}(S_R V_l^H) = \dim V_l^H$ different critical G-orbits of F with isotropy at least H. If H is maximal then these orbits must be homeomorphic to G/H. If $H = D_{2m}^d$ then $\dim V_l^H = [(l+m)/2m] = r$ provided $l/(2r+1) < m \leq l/(2r-1)$. Next one checks that D_{2m}^d is subconjugate to precisely $r - 1$ isotropy groups which must be of the form D_{2n}^d with $n \leq l$ and n/m an odd integer. If $m > l/9$ then $n > l/3$ so D_{2n}^d is a maximal isotropy group with one-dimensional fixed point space. Then there exist precisely $r - 1$ (critical) G-orbits in SV_l with isotropy greater than D_{2m}^d. Therefore one of the r critical G-orbits of F with isotropy at least D_{2m}^d must have isotropy equal to D_{2m}^d. For details see [Ba1]. □

Observe that in Theorem 8.7 it is allowed that the origin is not an isolated stationary orbit of φ_{λ_0}. Therefore we cannot take λ as parameter for the bifurcating stationary solutions. The results of this section are of course just a simple application of $\mathbb{Z}_2$-equivariant bifurcation theory to an $O(3)$-equivariant situation. An infinite-dimensional version of the $\mathbb{Z}_2$-equivariant bifurcation result needed for 8.7 is due to Böhme [Böh] and Marino [Ma]; see also [Ra2], §11. Similarly, Theorem 8.6 follows from a result of Fadell and Rabinowitz in [FaR1] provided the flow φ_λ satisfies the equation $\dot{x} = \lambda x - \nabla F(x)$ as in 8.7. Due to our general bifurcation theory in §7.5 it is not essential that λ enters only linearly. It is also not difficult to find examples where one can apply the results of §7.5 for $\mathbb{Z}_p$ instead of $\mathbb{Z}_2$.

8.8 Remark:

a) A detailed study of the bifurcation of stationary $O(3)$-orbits of

$$\dot{x} = f_\lambda(x) = \lambda x + o(\|x\|)$$

can be found in [CLM]. There it is not assumed that f_λ is a gradient. In that case one can only prove the existence of at least one branch of stationary solutions with maximal isotropy H provided $\dim V_l^H$ is odd. This is a simple consequence of the global bifurcation theorem of Rabinowitz [Ra1]. Using degree arguments it is also not difficult to see that there exist solution branches with isotropy equal to D_{2m}^d for every $3 \leq m \leq l$ if all solutions are normally hyperbolic. So these types of stationary solutions exist for a generic flow. Observe that for the special flows considered in Theorem 8.7 there exist stationary G-orbits with submaximal isotropy D^d_{2m} for $l/9 < m \leq l/3$ even without the hypothesis on normal hyperbolicity.

b) For $l = 3$ (see Figure 8.3) generically there is precisely one branch of stationary solutions with isotropy H for each maximal H, i.e. for $H = O(2)^-$, O^- and D^d_6. There are no branches with submaximal isotropy (D^z_2, D^z_3, $\mathbb{Z}^-_2$ or $\mathbb{Z}_1$); see [CLM], §6. The case $l = 5$ is somewhat surprising (see Figure 8.4). Generically there exist precisely one branch with maximal isotropy $O(2)^-$ (respectively D^d_{10}, D^d_8, D^d_6 and D^d_4), one with submaximal isotropy D^z_5 (respectively D^z_4) plus at least one with isotropy D^z_2. There are no stationary solutions with isotropy $D^z_3, D_2, \mathbb{Z}_2, \mathbb{Z}^-_2$ or $\mathbb{Z}_1$. Moreover, the branches with isotropies D^d_8 and D^d_4 are coupled; i.e. they bifurcate either both subcritically or both supercritically. The same is true with the submaximal branches D^z_5 and D^z_4. The proofs of these statements require the explicit knowledge of the $O(3)$-equivariant maps $V_l \to V_l$ up to third (or even fifth) order. It is not clear whether they can be proved using a more topologically minded approach without this knowledge.

8.4 Bifurcating connecting orbits

As in §8.3 we consider a continuous family of $O(3)$-equivariant gradient-like flows φ_λ, $\lambda \in \mathbb{R}$, on V_l for odd l. We suppose that the hypotheses (H1) and (H2) from §8.3 hold. In order to prove the existence of connecting orbits between the bifurcating solutions we assume one more regularity hypothesis.

(H3) For each maximal isotropy group H in $V_l - 0$ there exist precisely $\dim V^H_l$ branches of stationary solutions which can be parametrized in the form $[0, \epsilon] \ni s \mapsto (\lambda_i(s), u_i(s))$, $i = 1, \ldots, \dim V^H_l$. Here $\lambda_i(0) = \lambda_0$, $u(0) = 0$, $u_i(s) \in V^H_l - 0$ for $s > 0$ and λ_i is strictly monotone (decreasing or increasing).

8.10 Theorem:

Let $l = 3$ and suppose φ_λ satisfies (H1), (H2) and (H3). Moreover we assume that there are no bifurcating branches of stationary solutions with submaximal isotropy. These assumptions hold generically. We write S_λ for the bifurcating invariant set of stationary solutions and connecting orbits.

a) If all three branches with maximal isotropy bifurcate supercritically then for each $\lambda > \lambda_0$ near λ_0 the set S_λ attracts all orbits in a neighborhood of the origin (except the origin, of course). It separates V_3 into at least two components, the origin lying in a bounded one. In particular, $\dim S_\lambda \geq 6$. In addition to the three stationary G-orbits Gx_1, Gx_2, Gx_3 with maximal isotropy $G_{x_1} = O(2)^-$, $G_{x_2} = O^-$, $G_{x_3} = D^d_6$, the set S_λ contains flow orbits that connect Gx_1 with Gx_3, and Gx_2 with Gx_3, both with isotropy D^z_2, and flow orbits connecting Gx_1 with Gx_2, and Gx_3 with Gx_2 with isotropy D^z_3. An analoguous result holds if all stationary branches bifurcate subcritically.

b) If one of the stationary branches is subcritical and the other two are supercritical then for $\lambda > \lambda_0$ near λ_0 the two stationary G-orbits of S_λ must be connected by a flow orbit. Moreover, for these λ we have $\dim S_\lambda \geq 4$ and S_λ attracts an at *least five-dimensional subset of $V_3 - 0$. An analogous result holds if two branches bifurcate subcritically and one supercritically.*

Proof:
The group $\mathbb{Z}_2^c$ acts on $V_3 - 0$ without fixed points. We write ℓ for the $\mathbb{Z}_2^c$-length as usual. Since no stationary solutions bifurcate subcritically it follows that $\ell^u(0, \varphi_\lambda) = 0$ for $\lambda \leq \lambda_0$. Now $\ell^u(0, \varphi_\lambda) = \dim V_3 = 7$ for $\lambda > \lambda_0$, so by the theorems 7.10, 7.11 and their proofs

$$\ell(S_\lambda) = \ell\big(\mathcal{C}(S_\lambda)\big) = 7.$$

This implies $\dim S_\lambda \geq 6$. Moreover, S_λ contains the ω-limit set of a small sphere around 0.

In order to see the connecting orbits we apply Theorem 7.10 once more to the flow restricted to the two-dimensional fixed point spaces of D_2^z and D_3^z. We write $\mathrm{Fix}(H)$ for V_3^H. Then for $\lambda > \lambda_0$ we have $\ell\big(S_\lambda \cap \mathrm{Fix}(D_2^z)\big) = 2$. Now $\mathrm{Fix}(D_2^z)$ contains the one-dimensional fixed point spaces $\mathrm{Fix}\big(O(2)^-\big)$ and $\mathrm{Fix}(O^-)$. In addition, $\big(G \cdot \mathrm{Fix}(D_6^d)\big) \cap \mathrm{Fix}(D_2^z)$ consists of two one-dimensional subspaces $\mathrm{Fix}(D_6^d)$ and $\mathrm{Fix}(gD_6^d g^{-1})$ for some $g \in ND_2^z - ND_6^d$. Schematically the picture is as in Figure 8.11 (see [Lau]):

8.11 Figure:
The maximal fixed point spaces contained in the two-dimensional fixed point space $\mathrm{Fix}(D_2^z)$.

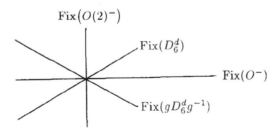

Observe that the stationary solutions $Gx_1 \cong G/O(2)^-$ and $Gx_2 \cong G/O^-$ intersect $\mathrm{Fix}(D_2^z)$ in two antipodal points whereas $Gx_3 \cong G/D_6^d$ intersects $\mathrm{Fix}(D_2^z)$ in four points (which form one WD_2^z-orbit because $WD_2^z \cong \mathbb{Z}_2 \times \mathbb{Z}_2$). Since we excluded stationary solutions with submaximal isotropy, $\ell(S_\lambda) = 2$ implies the existence of the connecting orbits as claimed. For this we need not quote the Poincaré-Bendixon theorem as in [Lau]. So our argument is not restricted to two dimensions. The situation in $\mathrm{Fix}(D_3^z)$ is the same with the roles of O^- and D_6^d exchanged. This proves part a).

For the proof of b) we need to know the $\mathbb{Z}_2^c$-length of $O(3)$-orbits. This is the content of the next proposition.

8.12 Proposition:

Let H be a closed subgroup of $O(3)$. Then

$$\ell\big(O(3)/H\big) = \begin{cases} 1 & \text{if } H \subset SO(3); \\ 2 & \text{if } H = \mathbb{Z}_{2m}^-; \\ 3 & \text{if } H = O(2)^-, O^-, D_{2m}^d, D_m^z; \\ \infty & \text{if } -1 \in H. \end{cases}$$

Postponing the proof of 8.12 to the end of this section we continue with the proof of 8.10 b). The length of the bifurcating stationary solutions Gx_i is 3 for every $i = 1, 2, 3$ according to 8.12. Thus for $\lambda < \lambda_0$ we have $\ell(S_\lambda) = 3$. Now Theorem 7.10 implies $\ell(S_\lambda) \geq 7 - 3 = 4$ if $\lambda > \lambda_0$. But the disjoint union of two sets of length 3 also has length 3 (see Theorem 4.15), so S_λ cannot just consist of the two stationary $O(3)$-orbits. Since (at least) one of the stationary solutions in S_λ, $\lambda > \lambda_0$, has a finite isotropy group the same is true for the isotropy on the connecting orbit. This implies $\dim S_\lambda \geq 4$. Finally, the proof of Theorem 7.11 shows that there exists a compact G-invariant subset A of $\{x \in V_3 : \|x\| = \epsilon\}$ for $\epsilon > 0$ small, with $S_\lambda = \omega(A)$ and $\ell(A) = 4$. We want to show that $\dim A \geq 4$ which implies that S_λ attracts the (at least) five-dimensional set $\varphi_\lambda(A, \mathbb{R})$. If $\dim A = 3$ then $\dim A/O(3) = 0$. Then we can cover A by open G-invariant neighborhoods U_i, $i \in I$, of G-orbits with $U_i \cap U_j = \emptyset$ if $i \neq j$. If the U_i are small enough we have $\ell(U_i) \leq 3$ according to Proposition 8.12. Then Theorem 4.15 again yields $\ell(A) \leq 3$, a contradiction. $\square$

One can even determine the isotropy of the connecting orbits in 8.10 b). Observe that the isotropy group of $\varphi_\lambda(x, t)$ is independent of t because $\varphi_\lambda(-, -t)$ is inverse to $\varphi_\lambda(-, t)$. For instance, suppose the two supercritical branches have isotropy $O(2)^-$ and O^-. It is easy to see that the connecting orbit cannot have isotropy D_2^z (just look at Figure 8.11). If there is no connecting orbit with isotropy D_3^z then one can show that $\ell(S_\lambda) = 3$, contradicting $\ell(S_\lambda) \geq 4$. We sketch such an argument in Remark 8.15. Looking at $\text{Fix}(D_3^z)$ one sees that $S_\lambda \cap \text{Fix}(D_3^z)$ attracts a two-dimensional "sector" of $\text{Fix}(D_3^z) - 0$. The group orbit of this sector is the five-dimensional subset of $V_3 - 0$ attracted by S_λ. (As mentioned above, the picture in $\text{Fix}(D_3^Z)$ looks schematically as in Figure 8.11 with O^- and D_6^d interchanged.)

Next we study the case $l = 5$. Since generically there are at least eight bifurcating branches of stationary solutions (cf. Remark 8.8), it is impossible to deal with all cases within the bounds of this chapter. Therefore we collect only a few simple results and discuss other features informally. We assume one additional hypothesis.

(H4) There exists precisely one branch with isotropy D_5^z and one with isotropy D_4^z. There are no branches with isotropy D_3^z or D_2.

This hypothesis holds generically according to [CLM], Theorem 8.1.

8.13 Theorem:

Let $l = 5$ and suppose φ_λ satisfies (H1) – (H4). Let $S_\lambda \subset V_5 - 0$ denote the bifurcating invariant set which is defined for λ near λ_0 ($S_{\lambda_0} = \emptyset$).

a) If S_λ contains stationary solutions with isotropies D_8^d and D_4^d then it contains a connecting orbit between these. The isotropy along the connecting orbit is D_2. The same is true with $(O(2)^-, D_6^d)$ instead of (D_8^d, D_4^d) and D_3^z instead of D_2.

b) If S_λ contains stationary solutions with isotropies $O(2)^-$, D_8^d and D_4^z then it contains two flow orbits connecting the D_4^z-solution with the $O(2)^-$-solution respectively the D_8^d-solution. The isotropy along the connecting orbits is D_4^z. The same is true with D_8^d replaced by D_{10}^d, and D_4^z by D_5^z.

Proof:

All statements can be proved by restricting the flow to the appropriate two-dimensional fixed point spaces $\mathrm{Fix}(H)$ for $H = D_2, D_3^z, D_4^z$ and D_5^z. The arguments are similar to those of Theorem 8.10.   □

Since generically the D_8^d- and the D_4^d-solutions are coupled (i.e. they bifurcate either both subcritically or both supercritically) the first part of 8.13a) always applies, either for $\lambda < \lambda_0$ or for $\lambda > \lambda_0$. It is also possible to compute the length of the invariant subsets of S_λ which consists of the stationary solutions plus the connecting orbits between these as in Theorem 8.13. For instance, the subset of S_λ described in 8.13a) is $S_\lambda^{(D_2)} := G \cdot \left(S_\lambda \cap \mathrm{Fix}(D_2) \right)$ and $\ell\left(S_\lambda^{(D_2)}\right) = 4$. In fact, $\ell\left(S_\lambda^{(H)}\right) = 4$ also for $H = D_3^z, D_4^z$ or D_5^z provided S_λ contains all stationary solutions with isotropy at least D_3^z, D_4^z or D_5^z. This is the case when all stationary solutions bifurcate supercritically, for example. In that case it is even true that

$$\ell\left(S_\lambda^{(D_2)} \cup S_\lambda^{(D_3^z)} \cup S_\lambda^{(D_4^z)} \cup S_\lambda^{(D_5^z)}\right) = 4.$$

However, according to Theorem 7.11 we have $\ell(S_\lambda) = 11$, so there must exist many more connecting orbits. Of course in this case S_λ attracts all orbits in a neighborhood of the origin and separates V_5 into at least two parts with the origin in a bounded component. But even if some of the stationary solutions bifurcate subcritically and others supercritically there must exist more connecting orbits than those of Theorem 8.13. A detailed analysis would require to distinguish between many cases depending on the isotropies of those branches which bifurcate supercritically (respectively subcritically). This is not the goal of our investigation. If one is interested in a concrete problem and if one knows the isotropies of the supercritical and subcritical branches, then our method allows to prove the existence of connecting orbits and to find their limit sets and the isotropy along the orbits.

For $l > 5$ the situation gets even more complicated. In addition to the bifurcating branches with maximal isotropy there exist generically branches with isotropy D_{2m}^d, for $m \geq 3$, see Remark 8.8a). Consider a maximal isotropy group of $V_l - 0$ and set $d(H) := \dim V_l^H$. If (H3) holds and if $\varphi_\lambda|V_l^H$ is a Morse-Smale flow then S_λ^H is a sphere of dimension $d_l(H) - 1$ for $\lambda < \lambda_0$ respectively $d_r(H) - 1$ for $\lambda > \lambda_0$, where $d_l(H), d_r(H) \geq 0$ satisfy $d_l(H) + d_r(H) \geq d(H)$; use Theorem 7.11 and Remark 6.4c) to see this. In this case the set $G \cdot S_\lambda^H = S_\lambda^{(H)}$ has length $d_l(H) + 2$ respectively $d_r(H) + 2$. A similar remark applies to the dihedral groups D_{2m}^d, $m \geq 2$, instead of a maximal isotropy group H provided there exists precisely one bifurcating branch with isotropy D_{2mk}^d for odd $k \geq 1$. These remarks are in fact true for arbitrary representation spaces V of $O(3)$ instead of V_l as long as $\mathbb{Z}_2^c$ acts on $V - 0$ without fixed points. Of course, one can look at all other fixed point spaces too. The bifurcation theorems of §7.5 apply, the length of the stationary orbits is known and the lengths of certain stationary orbits plus connecting orbits can be computed in many cases. But it is not clear how one can organize this mass of data.

8.14 Remark:

Our results on connecting orbits differ somewhat from those of Fiedler and Mischaikow [FiM] and Lauterbach [Lau], where the case $l = 3$ has been treated. The connecting orbits established in Theorem 8.10 have also been found in [FiM] and [Lau]. In addition, it is proved that the $O(2)^-$-branch of stationary solutions is always coupled with one of the two other branches. So if the $O(2)^-$-branch bifurcates subcritically then (at least) one of the two other branches must also bifurcate subcritically. Moreover, the various possible dimensions of the unstable manifolds of the stationary orbits are determined. The regularity assumptions in [FiM] and [Lau] are stronger than our hypotheses (H1) – (H4). In particular, the stationary orbits have to be normally hyperbolic in these papers. Finally, Lauterbach also finds the isotropy of the connecting orbits. This can be done with our method too as we shall indicate in 8.15 below. The method of Lauterbach consists of determining the possible equivariant polynomials $V_3 \to V_3$, at least up to degree 3 (or even 5). Knowing these he analyzes in detail the flow associated to a vector field which is given by these low order polynomials; see also [CLM]. But if l increases or if it is necessary to know the polynomials of degree 5 or higher this method becomes very complicated. The approach of Fiedler and Mischaikow is very close in spirit to the one of this chapter. Whereas we considered only the action of $\mathbb{Z}_2^c$ on V_3 and used the $\mathbb{Z}_2^c$-equivariant Conley index they forget the $O(3)$-action and work with the non-equivariant Conley index. The action of the full group $O(3)$ enters only in the way that the stationary solutions form manifolds homeomorphic to $O(3)/H$, where $H = O(2)^-, O^-, D_6^d$. An important tool in [FiM] is the connection matrix. This concept has not been developed in the equivariant context so far. Whereas we use to a certain extent the multiplicative structure of the equivariant cohomology of the $\mathbb{Z}/2$-Conley index, Fiedler and Mischaikow use the vector space structure of the homology of the non-equivariant Conley index. The method of the future would be to combine these two approaches and to develop for instance the connection matrix theory also in the equivariant setting.

An interesting feature of our approach to the existence of connecting orbits is that the orbits continue to exist if we perturb the flow in such a way that only the $\mathbb{Z}_2^c$-equivariance is maintained. Consider for example an isolated invariant set S with $\ell(S) = \ell(C(S)) = 4$ as in Theorem 8.10. It consists of two stationary orbits $Gu \cong G/O(2)^-$, $Gv \cong G/D_6^d$ and a connecting orbit with isotropy D_2^z. If we perturb the flow then we obtain an isolated invariant set S' which is only $\mathbb{Z}_2^c$-invariant. The stationary orbits Gu and Gv correspond to $\mathbb{Z}_2^c$-invariant submanifolds M_u and M_v of S', respectively, provided Gu and Gv are normally hyperbolic. By our results of §6.4 we know $\ell(S') \geq 4$ and $\ell(M_u) = 3 = \ell(M_v)$. So there must exist an orbit connecting M_u and M_v in S'. Due to normal hyperbolicity, M_u and M_v are $\mathbb{Z}_2^c$-homeomorphic to Gu and Gv (see [HPS]). This imposes also restrictions on the connecting orbit. In particular, it cannot be $\mathbb{Z}_2^c$-homeomorphic to $G/\mathbb{Z}_2^- \times \mathbb{R}$. This is not clear a priori because as a subset of V_3 the connecting orbit in $V_3^{D_2^z}$ can be perturbed to one lying in $V_3^{\mathbb{Z}_2^z}$.

We conclude this chapter with the proof of Proposition 8.12, i.e. with the computation of the lengths of the orbit spaces $O(3)/H$ considered as $\mathbb{Z}_2^c$-spaces. If H is contained in $SO(3)$ then $O(3)/H \cong (SO(3)/H) \times \mathbb{Z}_2^c$ hence $\ell(O(3)/H) = 1$ by the monotonicity of ℓ. And if H contains -1 then $\mathbb{Z}_2^c$ has a fixed point on $O(3)/H$, so $\ell(O(3)/H) = \infty$. It remains to consider the subgroups $H = O(2)^-, O^-, D_{2m}^d, D_m^z$ and $\mathbb{Z}_{2m}^-$. If $H = O(2)^-$ then one checks that $O(3)/H \cong SO(3)/SO(2) \cong S^2$ and the induced action of $\mathbb{Z}_2^c$ on S^2 is the antipodal action. Therefore $\ell(O(3)/H) = \ell(S^2) = 3$. Next we consider the case $H = \mathbb{Z}_{2m}^-$. The subspaces $(SO(2) \times \mathbb{Z}_2^c)/\mathbb{Z}_{2m}^-$ of $O(3)/\mathbb{Z}_{2m}^-$ are homeomorphic to S^1 with the antipodal action. Therefore

$$\ell(O(3)/\mathbb{Z}_{2m}^-) \geq \ell\left((SO(2) \times \mathbb{Z}_2^c)/\mathbb{Z}_{2m}^-\right) = \ell(S^1) = 2.$$

On the other hand, the cohomology of the orbit space $(O(3)/\mathbb{Z}_{2m}^-)/\mathbb{Z}_2^c$ with $\mathbb{F}_2$-coefficients is

$$H^*\left(O(3)/(\mathbb{Z}_{2m} \times \mathbb{Z}_2^c)\right) \cong H^*\left(SO(3)/\mathbb{Z}_{2m}\right) \cong H^*(S^3/\mathbb{Z}_{2m}^*)$$

where $\mathbb{Z}_{2m}^* \subset S^3$ is the inverse image of $\mathbb{Z}_{2m} \subset SO(3)$ under the two-sheeted covering map $S^3 \to SO(3)$. Below we use this notation also for other subgroups of $SO(3)$. Now $\mathbb{Z}_{2m}^* \cong \mathbb{Z}_{4m}$ and

$$H^i(S^3/\mathbb{Z}_{4m}) = \begin{cases} H^i(B\mathbb{Z}_{4m}) & \text{if } 0 \leq i \leq 3, \\ 0 & \text{if } i \geq 4. \end{cases}$$

The cohomology of $B\mathbb{Z}_{4m}$ is well known (see [FiP], §VI.2, for instance). We obtain:

$$H^*(S^3/\mathbb{Z}_{4m}) \cong \mathbb{F}_2[y] \otimes E[x]/y^2$$

where $\deg(x) = 1$ and $\deg(y) = 2$. Therefore $x^2 = 0$ which implies $\ell(O(3)/\mathbb{Z}_{2m}^-) \leq 2$.

Now consider $H = D_{2m}^d$. Since $\mathbb{Z}_{2m}^-$ is subconjugate to D_{2m}^d we have

$$\ell(O(3)/D_{2m}^d) \geq \ell(O(3)/\mathbb{Z}_{2m}^-) = 2.$$

Let $n = 2^k$ be the highest power of 2 which divides m. Then D_{2n}^d is subconjugate to D_{2m}^d, so $\ell(O(3)/D_{2m}^d) \geq \ell(O(3)/D_{2n}^d)$. But the $\mathbb{Z}_2^c$-map $O(3)/D_{2n}^d \to O(3)/D_{2m}^d$ induces a monomorphism on Borel cohomology because

$$(O(3)/D_{2m}^d)/\mathbb{Z}_2^c \cong O(3)/(D_{2m} \times \mathbb{Z}_2^c) \cong SO(3)/D_{2m} \cong S^3/D_{2m}^*$$

and D_{2n}^* is the 2-Sylow subgroup of D_{2m}^*. Therefore $\ell(O(3)/D_{2m}^d) = \ell(O(3)/D_{2n}^d)$. Again the cohomology of S^3/D_{2n}^* is known (see [FiP], §VI.5):

$$H^*(S^3/D_{2n}^*) \cong \mathbb{F}_2[x, y]/(x^2 + xy + y^2, x^2y + xy^2)$$

where $\deg(x) = \deg(y) = 1$. So the elements of $H^1(S^3/D_{2n}^*)$ are $x, y, x+y$ (and 0). It is clear that x^2, y^2 and $(x+y)^2 = x^2 + y^2$ are not zero but x^3, y^3 and $(x+y)^3$ are zero $(x^3 = x^2y + xy^2 = y^3, (x+y)^3 = x^3 + x^2y + yx^2 + y^3)$. This implies $\ell(O(3)/D_{2m}^d) = 3$.

The argument for $H = D_m^z$ is the same. We are led to compute the cohomology of S^3/D_m^* for $m \geq 2$ which is again $\mathbb{F}_2[x,y]/(x^2+xy+y^2, x^2y+xy^2)$. Finally we have to compute the length of $O(3)/O^-$. But this is the same as $\ell(O(3)/D_4^d)$ because D_4^d is the 2-Sylow subgroup of O^-. So $\ell(O(3)/D_m^z) = 3 = \ell(O(3)/O^-)$.   □

8.15 Remark:

For further investigations it is important to know the length of invariant sets consisting of stationary orbits and connecting orbits. We want to sketch such a computation. Suppose the invariant set S consists of two stationary orbits Gu, Gv and one connecting orbit between these; here $G = O(3)$, of course. We assume $G_u = O(2)^-$ and $G_v = D_6^d$ and the isotropy along the connecting orbit is $\mathbb{Z}_2^-$, so $S - (Gu \cup Gv) \cong G/\mathbb{Z}_2^- \times \mathbb{R}$. We know that the length of Gu and Gv is 3 and we want to show that $\ell(S) = 3$. Thus the additional connecting orbit does not increase the length. This depends heavily on the isotropies. If the isotropy along the connecting orbit is D_2^z, for instance, then the length would be 4. Results of this type can be used to determine the isotropy of connecting orbits in certain situations. The example which we chose is of interest in the case $l = 3$; see Theorem 8.10. To compute $\ell(S)$ set $A := S - Gv$, $B := S - Gu$ and consider the Mayer-Vietoris sequence of the $\mathbb{Z}_2^c$-triad $(S; A, B)$. We need the following part of this sequence (h denotes the Borel cohomology with coefficients in $\mathbb{F}_2$ as in Example 4.4):

$$h^2(A \cap B) \xrightarrow{\delta} h^3(S) \xrightarrow{\alpha} h^3(A) \oplus h^3(B) \xrightarrow{\beta} h^3(A \cap B) \xrightarrow{\delta} h^4(S).$$

We have to show that $w^3 \cdot 1_S = 0 \in h^3(S)$ where $w \in h^1(\text{pt})$ and $1_S \in h^0(S)$ as usual. Since $\ell(A) = 3 = \ell(B)$ we have $\alpha(w^3 \cdot 1_S) = (w^3 \cdot 1_A, w^3 \cdot 1_B) = (0,0)$. Hence, we are done if we can show that $\delta(y) \neq w^3 \cdot 1_S$ for any non-zero element $y \in h^2(A \cap B)$. Now $A \cap B \cong G/\mathbb{Z}_2^- \times \mathbb{R} \simeq G/\mathbb{Z}_2^-$ and the cohomology of $G/\mathbb{Z}_2^-$ is well known:

$$h^*(A \cap B) \cong h^*(G/\mathbb{Z}_2^-) \cong \mathbb{F}_2[y] \otimes E[x]/y^2$$

where $\deg(x) = 1$ and $\deg(y) = 2$; cf. the proof of Proposition 8.12. So there is only one non-zero element $y \in h^2(A \cap B)$. Since $\ell(A \cap B) = 2$ we obtain $w \cdot 1_{A \cap B} = x \in h^1(A \cap B)$. This implies $w \cdot y = x \smile y \neq 0 \in h^3(A \cap B)$. We claim that $w \cdot y$ is not in the image of β and consequently $\delta(w \cdot y) \neq 0 \in h^4(S)$. First of all $h^3(A) = 0$ because $h^3(A) \cong h^3(O(3)/O(2)^-) \cong h^3(S^2) = 0$. Next the inclusion $A \cap B \hookrightarrow A$ is up to homotopy the map $G/\mathbb{Z}_2^- \to G/D_6^d$. And $w \cdot y$ cannot lie in the image of $h^3(G/D_6^d) \to h^3(G/\mathbb{Z}_2^-)$ because all elements of $h^3(G/D_6^d)$ are products of three elements of $h^1(G/D_6^d)$ but $w \cdot y$ is not a product of three elements of $h^1(G/\mathbb{Z}_2^-)$; for this argument compare the proof of Proposition 8.12 once more. Thus we have proved that $\delta(w \cdot y) \neq 0 \in h^4(S)$. If $\delta(y) = w^3 \cdot 1_S$ then $w^4 \cdot 1_S = \delta(w \cdot y) \neq 0$. But this is not possible as a consequence of the duality theorem 4.17. The set S can be considered as a subspace of the sphere SV_3. In fact, it is contained in the complement of an orbit of the form G/O^-: Theorem 4.17 implies $\ell(S) \leq \ell(SV_3) - \ell(G/O^-) = 7 - 3 = 4$, hence $w^4 \cdot 1_S = 0$.

Chapter 9

Multiple periodic solutions near equilibria
of symmetric Hamiltonian systems

9.1 Introduction

In this chapter we study a very different situation from the one investigated in chapter 8. Let $H: \mathbb{R}^{2n} \to \mathbb{R}$ be C^2 and consider the Hamiltonian system

$$(HS) \qquad\qquad \dot{z} = J \nabla H(z)$$

on $\mathbb{R}^{2n}$ where $J = \begin{pmatrix} 0 & -\mathrm{Id} \\ \mathrm{Id} & 0 \end{pmatrix}$ is the usual symplectic structure. Suppose that the compact Lie group G acts (generalized) symplectically on $\mathbb{R}^{2n}$ and that H is G-invariant. Moreover we assume that $\nabla H(0) = 0$ so that $z \equiv 0$ is a stationary solution of (HS). This is automatic if the origin is the only fixed point of the action. We are interested in periodic solutions of (HS) in a neighborhood of the origin. Given a solution of (HS) we call the image $\tau = z(\mathbb{R})$ a trajectory. A closed trajectory is the image of a periodic solution of (HS). Due to our assumptions on H and on the action of G any trajectory τ of (HS) gives rise to a G-orbit $G\tau$ of trajectories. We want to count the number of G-orbits of closed trajectories near the origin.

This is an old problem. Classical results in this direction (without symmetry) are due to Lyapunov [Ly] and, more recently, to Weinstein [We], Moser [Mos] and Fadell and Rabinowitz [FaR2]. We shall generalize these results to the symmetric situation. An equivariant version of the theorem of Weinstein and Moser is due to Montaldi, Roberts and Stewart [MoRS]. We are able to deal with more general actions than those considered in [MoRS] and, in addition, improve the result of [MoRS] even for their actions.

Clearly, H is a first integral of (HS). Therefore one may try to find closed trajectories on energy surfaces $H^{-1}(c)$ with c near $H(0)$. This will be done in §9.2 where we generalize the Weinstein-Moser theorem to the symmetric situation. There we also recall a variational reformulation of our problem. Another way to count the closed trajectories near the origin is to take the period as a parameter. It is well known that the presence of a purely imaginary eigenvalue $i\beta$ of the linearization $JH''(0)$ is a necessary condition for the existence of periodic solutions near the origin. The linearized equation has periodic solutions with period $2\pi/\beta$, so one can try to find periodic solutions of (HS) near 0 with the period near $2\pi/\beta$ as parameter. This is the goal of §9.3 where we generalize the main result of [FaR2] to the symmetric case.

Given a closed trajectory τ let $G_\tau = \{g \in G : g\tau = \tau\}$ be the isotropy group of τ. Knowing the isotropy groups of closed trajectories of the linearized equation one may look at closed trajectories of (HS) having the same isotropy. This can be achieved using the usual trick of restriction to fixed point spaces. We pursue this approach in §9.4 and study some special cases. In particular, we prove the existence of brake orbits and normal mode solutions near the origin which appear for certain symmetries.

9.2 Fixed energy

As in the introduction to this chapter let $H \colon \mathbb{R}^{2n} \to \mathbb{R}$ be a C^2-map with $H(0) = 0$ and $\nabla H(0) = 0$. Then $z \equiv 0$ is a stationary solution of the Hamiltonian system

$$(HS) \qquad\qquad \dot{z} = J\nabla H(z).$$

We are interested in periodic solutions of (HS) on energy surfaces $H^{-1}(\varepsilon)$ for ε close to 0. We assume the following hypotheses on H.

($H1$) $H \in C^2(\mathbb{R}^{2n}, \mathbb{R})$ and $H(0) = 0$, $\nabla H(0) = 0$, $H''(0) \in \mathrm{GL}(2n)$, and $JH''(0)$ has a purely imaginary eigenvalue $i\beta$ with $\beta > 0$.

If $(H1)$ holds then the linearized Hamiltonian system

$$(LHS) \qquad\qquad \dot{v} = JH''(0)v$$

has a periodic solution with minimal period $2\pi/\beta$. Let $E \subset \mathbb{R}^{2n}$ be the sum of the generalized eigenspaces of $JH''(0)$ associated to the eigenvalues of the form $\pm ik\beta$, $k \in \mathbb{N}$. The dimension of E is even and E is invariant under the flow associated to (LHS). We also need the following hypothesis.

($H2$) $H''(0)|E$ is positive definite.

If $(H2)$ holds then $JH''(0)|E$ is diagonalizable. Every solution of (LHS) with initial value in E has (not necessarily minimal) period $2\pi/\beta$ and no solution with initial value in $\mathbb{R}^{2n} - E$ has this period. Observe that (LHS) has $\frac{1}{2} \dim E$ linearly independent periodic solutions with period $2\pi/\beta$.

To formulate the symmetry condition on H recall that a $2n \times 2n$-matrix A is said to be *symplectic* if $A^t J A = J$ and to be *antisymplectic* if $A^t J A = -J$. We call A *generalized symplectic* if A is either symplectic or antisymplectic and write $\mathrm{GSp}(2n)$ for the set of all generalized symplectic matrices. This is a closed (but not compact) subgroup of $\mathrm{GL}(2n)$ and $|\det A| = 1$ for $A \in \mathrm{GSp}(2n)$. Note that $A \in \mathrm{GSp}(2n)$ implies $A^t, A^{-1} \in \mathrm{GSp}(2n)$.

The map $\sigma \colon \mathrm{GSp}(2n) \to \{\pm 1\}$ defined by $A^t J A = \sigma(A)J$ is a homomorphism and satisfies $\sigma(A) = \sigma(A^t) = \sigma(A^{-1})$. Clearly, $\ker \sigma = \mathrm{Sp}(2n)$ is the group of symplectic matrices. Setting $I_1 = \begin{pmatrix} -\mathrm{Id} & 0 \\ 0 & \mathrm{Id} \end{pmatrix}$ the map $A \mapsto I_1 A$ maps $\mathrm{Sp}(2n)$ to $\mathrm{GSp}(2n) - \mathrm{Sp}(2n)$ and vice versa. Finally, a representation $\rho \colon G \to \mathrm{GL}(2n)$ of the group G on $\mathbb{R}^{2n}$ is said to be generalized symplectic if $\rho(G) \subset \mathrm{GSp}(2n)$. As usual we write gz for $\rho(g)z$ and $\sigma(g)$ for $\sigma(\rho(g))$. Since the groups we deal with are topological groups and the representations are continuous σ factors through the factor group G/G_0 where G_0 is the connected component of the unit of G:

$\sigma: G/G_0 \rightarrow \{\pm 1\}$. In particular, an action of a connected group is generalized symplectic iff it is symplectic, that is $\rho(G) \subset \mathrm{Sp}(2n)$.

The symmetry condition on H is as follows.

(H3) $H: \mathbb{R}^{2n} \rightarrow \mathbb{R}$ is invariant under a generalized symplectic action of a compact Lie group G on $\mathbb{R}^{2n}$.

9.1 Example:

a) If $\rho_0: G \rightarrow \mathrm{O}(n)$ is an orthogonal representation of G on $\mathbb{R}^n$ then

$$\rho: G \rightarrow \mathrm{GSp}(2n) \cap \mathrm{O}(2n), \quad g \mapsto \begin{pmatrix} \rho_0(g) & 0 \\ 0 & \rho_0(g) \end{pmatrix}$$

defines a symplectic representation of G on $\mathbb{R}^{2n}$. Specific examples of Hamiltonian systems which have this symmetry can be found in [MoRS], Examples 1.3. These include the spherical pendulum where $G = \mathrm{O}(2)$ and the Hénon-Heiles Hamiltonian where G is the dihedral group of order 6, for instance.

b) Set $G = \{1, I_1\} \cong \mathbb{Z}/2$ with $I_1 = \begin{pmatrix} -\mathrm{Id} & 0 \\ 0 & \mathrm{Id} \end{pmatrix}$ as above. This action is generalized symplectic and $\sigma(I_1) = -1$. Every classical Hamiltonian function $H(p,q) = \frac{1}{2}\|p\|^2 + V(q)$ is invariant with respect to this action. Hamiltonian systems with this symmetry have been studied by Rabinowitz [Ra3] and Szulkin [Sz2].

c) Set $G = \{1, I_1, I_2, I_1 I_2\} \cong \mathbb{Z}/2 \times \mathbb{Z}/2$ with I_1 as above and $I_2 = -I_1$. This action is generalized symplectic and $\sigma(I_1) = \sigma(I_2) = -1$. If H is a classical Hamiltonian function with even potential energy $V: \mathbb{R}^n \rightarrow \mathbb{R}$ then H is invariant with respect to this action. Convex Hamiltonian systems with this symmetry have been studied by van Groesen [vGr].

It is easy to check that the vector field $J\nabla H$ on $\mathbb{R}^{2n}$ satisfies

$$J\nabla H(gz) = \sigma(g)gJ\nabla H(z)$$

so it is "generalized equivariant". It follows immediately that $gz(\sigma(g)t)$ solves (HS) if z does. Thus each trajectory τ of (HS) gives rise to a G-orbit $G\tau$ of trajectories of (HS). We want to count the number of G-orbits of closed trajectories of (HS) on energy surfaces $H^{-1}(\varepsilon)$ with ε near 0. Observe that E is G-invariant.

9.2 Theorem:
If H satisfies (H1), (H2) and (H3) then (HS) has at least

$$\langle (\dim E)/2(1 + \dim G - \mathrm{rk}\, G) \rangle$$

different G-orbits of closed trajectories on $H^{-1}(\varepsilon)$ for sufficiently small $\varepsilon > 0$. Here $\mathrm{rk}\, G$ is the dimension of the maximal torus of G and $\langle x \rangle$ denotes the least integer greater than or equal to x. The periodic solutions corresponding to these closed trajectories have periods near $2\pi/\beta$.

If G is trivial then Theorem 9.2 is due to Weinstein [We] and Moser [Mos]. As we shall see at the end of the proof the estimate can be improved if one knows more about the action of G on E and if one can calculate the torus category of certain orbit spaces. Moreover, 9.2 can be improved if one takes the symmetry of the periodic solutions into account. This simple generalization will be investigated in §9.4 where Theorem 9.2 appears only as a special case.

It is clear that Theorem 9.2 extends to the slightly more general situation where $\mathbb{R}^{2n}$ is replaced by a manifold M with symplectic structure Ω. A group G acts generalized symplectically on M if for each $g \in G$ the map $\mu_g: M \ni z \mapsto gz \in M$ is C^1 and $\mu_g^* \Omega = \pm\Omega$. If $z \in M$ is a fixed point of such an action then $T_z M$ is a representation space of G and the action of G on $T_z M$ is generalized symplectic. Let $H: M \to \mathbb{R}$ be a G-invariant C^2-map and consider the Hamiltonian system $\dot{z} = X_H(z)$ where the vector field X_H is defined by $\Omega\big(X_H(z), u\big) = DH(z)u$ for $z \in M$ and $u \in T_z M$. If $z_0 \in M$ is a zero of X_H and a fixed point of the action of G then one can reduce the existence of periodic solutions near the stationary solution z_0 to Theorem 9.2 using an equivariant version of Darboux's theorem.

Proof of Theorem 9.2:

First we recall the variational formulation of the problem. Setting $s = 2\pi t/\lambda$ equation (HS) becomes

$$(HS_\lambda) \qquad\qquad \dot{z} = \frac{\lambda}{2\pi} J \nabla H(z)$$

where the dot now denotes d/ds. Periodic solutions of (HS) with period λ correspond to 2π-periodic solutions of (HS_λ). Let $Z = W^{1,2}(S^1, \mathbb{R}^{2n})$ be the real Hilbert space of 2π-periodic functions $z: \mathbb{R} \to \mathbb{R}^{2n}$ with square integrable derivative. The scalar product on Z is given by

$$\langle z_1, z_2 \rangle_Z = \frac{1}{2\pi} \int_0^{2\pi} \big(z_1(s) \cdot z_2(s) + \dot{z}_1(s) \cdot \dot{z}_2(s)\big)\, ds$$

where $\cdot$ denotes the usual scalar product in $\mathbb{R}^{2n}$. Consider the functionals

$$A: Z \longrightarrow \mathbb{R}, \quad A(z) := \frac{1}{2} \int_0^{2\pi} J\dot{z}(s) \cdot z(s)\, ds,$$

and

$$B: Z \longrightarrow \mathbb{R}, \quad B(z) := \frac{1}{2\pi} \int_0^{2\pi} H\big(z(s)\big)\, ds.$$

Both functionals are of class C^2. It is straightforward to check that the critical points of $F_\lambda := A - \lambda B$ solve (HS_λ). In fact,

$$\langle \nabla F_\lambda(z), v \rangle_Z = F_\lambda'(z)v = \int_0^{2\pi} \left(J\dot{z}(s) + \frac{\lambda}{2\pi}\nabla H\big(z(s)\big)\right) \cdot v(s)\, ds.$$

Moreover, since $H\big(z(s)\big)$ is independent of s if z solves (HS), the critical points of F_λ satisfying $B(z) = \varepsilon$ are precisely the periodic solutions of (HS) with energy ε. This suggests to look for critical points of $A|B^{-1}(\varepsilon)$. But due to the indefiniteness of A it is easier to consider the situation as a bifurcation problem and to make a finite-dimensional reduction than as a constrained variational problem.

Now we study how the invariance of H under G as formulated in $(H3)$ comes into play. Let $\sigma\colon G \to \{\pm 1\}$ be the sign function of the generalized symplectic representation of G on $\mathbb{R}^{2n}$, that is $g^t J g = \sigma(g) J$. The group $\{\pm 1\}$ is isomorphic to the automorphism group of $S^1 = \mathbb{R}/2\pi\mathbb{Z}$. In the following we do not distinguish between $\theta \in \mathbb{R}$ and the corresponding element of S^1. Then $\mathrm{Aut}(S^1) = \{\pm\mathrm{id}\}$ where $-\mathrm{id}(\theta) = -\theta$, of course. Let $\Gamma = S^1 \rtimes G$ be the semidirect product of S^1 and G under σ. The elements of Γ are pairs $\gamma = (\theta, g) \in S^1 \times G$ and the multiplication is given by

$$(\theta, g) \cdot (\theta', g') = \big(\theta + \sigma(g)\theta', gg'\big).$$

If σ is trivial then $\Gamma = S^1 \times G$. In general, Γ need not be commutative even if G is. The compact Lie group Γ acts on Z as follows. If $\gamma = (\theta, g) \in \Gamma$ and $z \in Z$ then $\gamma z \in Z$ is given by

$$(\gamma z)(s) = gz\big(\sigma(\gamma)(s + \theta)\big).$$

We leave the simple verification of $(\gamma\gamma')z = \gamma(\gamma' z)$ to the reader.

Next one sees that A is Γ-invariant because $g^t J g = \sigma(g) \cdot J$ for every $g \in G$. The Γ-invariance of B is an immediate consequence of the G-invariance of H. Therefore the G-orbits of closed trajectories of (HS) are in 1-1 correspondence with the Γ-orbits of critical points of F_λ.

The linearization of $\nabla F_\lambda(z) = \nabla A(z) - \lambda \nabla B(z) = 0$ at $z = 0$ is given by

$$F_\lambda''(0)v\colon Z \to \mathbb{R}, \quad w \mapsto \int_0^{2\pi} \Big(J\dot{v}(s) + \frac{\lambda}{2\pi} H''(0)v(s)\Big) \cdot w(s)\, ds.$$

Thus the kernel of $F_\lambda''(0)$ consists of all $v \in Z$ which solve the linear Hamiltonian system

(LHS_λ) $$\dot{v} = \frac{\lambda}{2\pi} J H''(0)v.$$

Setting $\lambda_0 = 2\pi/\beta$ the kernel $V := \ker F_{\lambda_0}''(0) \subset Z$ is isomorphic to $E \subset \mathbb{R}^{2n}$ where E is defined above after $(H1)$. Since F_λ is Γ-invariant and 0 is fixed by the action of Γ the gradient ∇F_λ and its linearization $F_\lambda''(0)$ at 0 are Γ-equivariant maps $Z \to Z$. Therefore V is Γ-invariant.

Now one can reduce the problem of finding solutions of $\nabla F_\lambda(z) = 0$ with $z \in B^{-1}(\varepsilon)$ and λ near λ_0 to a corresponding problem in V via a Lyapunov-Schmidt reduction. We refer the reader to [MaW], §6.5, for a detailed description. One finally obtains a C^1-function

$$f\colon B_\delta(V) = \{v \in V\colon \|v\| \le \delta\} \to V^\perp \subset Z$$

and a C^1-function $\alpha\colon B_\delta(V) \to \mathbb{R}$ such that

— $Z_\varepsilon := \{v \in V\colon B\big(v + f(v)\big) = \varepsilon\}$ is radially diffeomorphic to the unit sphere in V for $\varepsilon > 0$ small enough and
— the critical points v of $Z_\varepsilon \xrightarrow{\alpha} \mathbb{R}$ correspond to solutions $z = v + f(v)$ of $\nabla F_\lambda(z) = 0$ with $B(z) = \varepsilon$ where λ is near λ_0.

The construction in [MaW] is automatically Γ-equivariant. In conclusion, the number of G-orbits of periodic solutions of (HS) with energy $\varepsilon > 0$ sufficiently small and period near $2\pi/\beta$ is the same as the number of critical Γ-orbits of a C^1-function $\alpha \colon Z_\varepsilon \to \mathbb{R}$ where Z_ε is radially diffeomorphic to the sphere SV of V. We claim that the maximal torus $T\Gamma = S^1 \times T$ acts on SV without fixed points. This follows from the fact that $\gamma = (\theta, 1) \in S^1 \times T$ acts on $v \in SV$ via $(\gamma v)(s) = v(s + \theta)$; and the elements of $V - 0$ are not constant because of hypothesis $(H1)$.

Set $\mathcal{A} = \mathcal{A}_{SV} = \{\Gamma/\Gamma_v \colon v \in SV\}$ and let $\widehat{\mathcal{A}}$ be the set of all finite disjoint unions of elements of $\mathcal{A}$ as in §2.4. According to Theorem 2.19 α has at least $\widehat{\mathcal{A}}$-cat$(Z_\varepsilon) = \widehat{\mathcal{A}}$-cat$(SV)$ critical Γ-orbits. And Corollary 2.21 gives

$$\widehat{\mathcal{A}}\text{-cat}(SV) \geq (\dim V)/2(1 + \dim \Gamma - \operatorname{rk} \Gamma) = (\dim V)/2(1 + \dim G - \operatorname{rk} G).$$

□

9.3 Remark:

The estimate of Theorem 9.2 can be improved in the following way. Set

$$\mathcal{B} := \{T\Gamma/\Delta \colon \Delta \text{ is a closed subgroup of } T\Gamma \text{ and } \Delta \cap S^1 \times \{1\} \text{ is finite}\}.$$

Then

$$\widehat{\mathcal{A}}\text{-cat}(Z_\varepsilon) \geq (\dim V)/2 \max\{\widehat{\mathcal{B}}\text{-cat}(A) \colon A \in \widehat{\mathcal{A}}\}.$$

Proposition 2.13b) tells us that

$$\widehat{\mathcal{B}}\text{-cat}(A) \leq 1 + \dim(A/T\Gamma) \leq 1 + \dim \Gamma - \operatorname{rk} \Gamma$$

for every $A \in \widehat{\mathcal{A}}$. But in certain cases $\max\{\widehat{\mathcal{B}}\text{-cat}(A) \colon A \in \widehat{\mathcal{A}}\}$ can be smaller than $1 + \dim \Gamma - \operatorname{rk} \Gamma$.

One can also work with the length ℓ for the maximal torus $T\Gamma$ defined in Example 4.3. We know from Theorem 5.1 that $\ell(SV) = \frac{1}{2} \dim V$, so that a lower bound for the number of critical Γ-orbits of $\alpha \colon SV \to \mathbb{R}$ is given by

$$(\dim V)/2 \max\{\ell(\Gamma/\Gamma_v) \colon v \in SV\}.$$

According to Proposition 5.14 we have $\ell(\Gamma/\Gamma_v) \leq |N_{G_0} T/T| \cdot |G/G_0|$ where G_0 is the component of the unit in G. For $G = SO(3)$ or $G = SU(2)$ we have $|N_{G_0} T/T| = 2$ but $1 + \dim G - \operatorname{rk} G = 3$. Thus for these groups we obtain at least $\frac{1}{4} \dim V = \frac{1}{4} \dim E$ G-orbits of closed trajectories in $H^{-1}(\varepsilon)$ for sufficiently small $\varepsilon > 0$. We shall see in §9.4 that even this result can be improved.

9.4 Remark:

If $H''(0)$ is positive definite then there exist at least $\langle n/(1 + \dim G - \mathrm{rk}\, G) \rangle$ G-orbits of closed trajectories on $H^{-1}(\varepsilon)$ for sufficiently small $\varepsilon > 0$. This can be seen by splitting $\mathbb{R}^{2n}$ into the direct sum $\mathbb{R}^{2n} \cong E_1 \oplus \ldots \oplus E_r$ where E_j is the generalized eigenspace of $JH''(0)$ associated to $\pm i k \beta_j$, $k \in \mathbb{N}$, and where the $\beta_1, \ldots, \beta_r > 0$ are incommensurable. Here we used the fact that the positive definiteness of $H''(0)$ implies that $JH''(0)$ has only purely imaginary eigenvalues and is diagonalizable; see [We], [Mos]. There exists also a global version of this result. Let c be a regular value of H such that $H^{-1}(c)$ is compact. Set $R := \max\{\|z\| : z \in H^{-1}(c)\}$ and $r := \min\{z \cdot \nabla H(z)/\|\nabla H(z)\| : z \in H^{-1}(c)\}$. If $R^2 < 2r^2$ then (HS) has at least $\langle n/(1 + \dim G - \mathrm{rk}\, G) \rangle$ G-orbits of closed trajectories on $H^{-1}(c)$. The assumption $R^2 < 2r^2$ implies that $H^{-1}(c)$ bounds a strictly starshaped compact neighborhood of 0 in $\mathbb{R}^{2n}$. It is satisfied in the local setting if $JH''(0)$ has only purely imaginary eigenvalues $\pm i\beta_1, \ldots, \pm i\beta_n$ and $\beta_j/\beta_k < 2$ for all $j, k = 1, \ldots, n$. In order to prove this global version of Theorem 9.2 one has to develop a critical point theory for the strongly indefinite Γ-equivariant action functional $z \mapsto \int_0^{2\pi} J\dot{z}(s) \cdot z(s)\, ds$. This has been done in [BC2] where results of [BLMR], [Gi], [vGr] and [Sz2] have been generalized.

9.3 Fixed period

We continue to study the Hamiltonian system

$$(HS) \qquad\qquad \dot{z} = J\nabla H(z).$$

This time we want to find the number of periodic solutions near a stationary solution with the period and not the energy fixed. We shall generalize the result of Fadell and Rabinowitz [FaR2] to the symmetric situation. So we keep the assumption $(H3)$, that is $H : \mathbb{R}^{2n} \to \mathbb{R}$ is invariant under a generalized symplectic action of a compact Lie group G on $\mathbb{R}^{2n}$. We also keep the hypothesis $(H1)$, so that $H \in C^2(\mathbb{R}^{2n}, \mathbb{R})$ satisfies $H(0) = 0$, $\nabla H(0) = 0$, $H''(0) \in \mathrm{GL}(2n)$ and $JH''(0)$ has a purely imaginary eigenvalue $i\beta$. As in §9.2 let $E \subset \mathbb{R}^{2n}$ be the generalized eigenspace of $JH''(0)$ associated to those eigenvalues which are integer multiples of $i\beta$. Hypothesis $(H2)$ can be relaxed to

$(H2')$ $JH''(0)|E$ is semisimple.

This is equivalent to the statement that the linearized Hamiltonian system

$$(LHS) \qquad\qquad \dot{v} = JH''(0)v$$

has $\frac{1}{2} \dim E$ linearly independent periodic solutions with (not necessarily minimal) period $2\pi/\beta$. Therefore the group $S^1 = \mathbb{R}/2\pi\mathbb{Z}$ acts on E as follows. Given $\theta \in S^1$ and $u \in E$ set $\theta \cdot u := v(\theta/\beta)$ where v solves (LHS) and $v(0) = u$. Since E is also G-invariant we obtain an action of $\Gamma = S^1 \rtimes G$ on E. Clearly, the isomorphism $v \mapsto v(0)$ between $V \subset W^{1,2}(S^1, \mathbb{R}^{2n})$ and E is Γ-equivariant. Here V is defined as in §9.2. The quadratic form

$$E \ni u \mapsto H''(0)u \cdot u \in \mathbb{R}$$

is invariant under this action and, hence, so are the eigenspaces E^+ respectively E^- on which this form is positive respectively negative definite. Moreover, $E \cong E^+ \oplus E^-$ and E^+, E^- are even-dimensional because $H''(0)$ is nonsingular so that the action of S^1 on E is fixed point free. We write $\operatorname{sgn}(H''(0)|E) = \dim E^+ - \dim E^-$ for the *signature* of this quadratic form.

In order to state our result recall the length ℓ of Example 4.3 for the maximal torus $T\Gamma$ of Γ. Set

$$\mu := \max\{\ell(\Gamma/\Gamma_u) : u \in E - 0\}.$$

9.5 Theorem:
If H satisfies $(H1)$, $(H2')$ and $(H3)$ then one of the following holds:

(i) *There exists a sequence z_i of periodic solutions of (HS) with period $2\pi/\beta$ and $\|z_i\| \to 0$ as $i \to \infty$.*

(ii) *There exist integers $d_l, d_r \geq 0$ with $d_l + d_r \geq |\operatorname{sgn}(H''(0)|E)/2\mu|$ and having the property: For each $\lambda < 2\pi/\beta$ (respectively $\lambda > 2\pi/\beta$) and λ sufficiently close to $2\pi/\beta$ there exist at least d_l (respectively d_r) G-orbits of closed trajectories of (HS) with period λ. These trajectories converge towards the origin as $\lambda \to 2\pi/\beta$.*

As mentioned above, for trivial G the result is due to Fadell and Rabinowitz. In this case $\Gamma = S^1$ and $\mu = 1$. If the signature $\dim E^+ - \dim E^-$ of the quadratic form on E is 0 then Theorem 9.5 is trivial because (ii) applies with $d_l = d_r = 0$. An example due to Moser [Mos] shows that there need not exist any closed trajectories near the equilibrium if $\operatorname{sgn}(H''(0)|E) = 0$. The result can be improved by taking the symmetry of the periodic solutions into account. As with Theorem 9.2 this will be deferred to the next section.

The results of the chapters 4 and 5 yield some upper bounds for μ. Suppose $G = T \times F$ where T is a torus (possibly $T = \{e\}$) and F is a finite group. If G acts symplectically on $\mathbb{R}^{2n}$ then $\mu = 1$ as a consequence of the normalization property. This is optimal. We also know that $\mu \leq |N_{G_0}T/T| \cdot |G/G_0|$; see Proposition 5.15. Thus for $G = SO(3)$ or $SU(2)$ we obtain $\mu \leq 2$ because the Weyl group is $\mathbb{Z}/2$ in these cases. It is not clear whether one can prove

$$d_l + d_r \geq |\dim E^+ - \dim E^-|/2(1 + \dim G - \operatorname{rk} G)$$

as one would expect from Theorem 9.2. But one can show that

$$d_l + d_r \geq |\dim E^+ - \dim E^-|/2(1 + \dim G).$$

To see this one replaces the length for $T\Gamma$ by the length for $S^1 = S^1 \times \{1\} \subset T\Gamma$ and uses Remark 5.16b). Probably one can even show that $\mu \leq 1 + \dim G$ but the inequality $\mu \leq 1 + \dim G - \operatorname{rk} G$ seems to be wrong.

Proof of Theorem 9.5:

We work in a slightly different setting than in the proof of 9.2. Let $Z = W^{1,2}(S^1, \mathbb{R}^{2n})$, $Y = L^2(S^1, \mathbb{R}^{2n})$ and $f_\lambda: Z \to Y$, $f_\lambda(z) := J\dot{z} + \frac{\lambda}{2\pi}\nabla H(z)$. Periodic solutions of (HS) with period λ correspond to zeroes of f_λ. Observe that $\langle \nabla F_\lambda(z), v \rangle_Z = \langle f_\lambda(z), v \rangle_Y$ where

$$F_\lambda(z) = A(z) - \lambda B(z) = \frac{1}{2}\int_0^{2\pi} J\dot{z}(s) \cdot z(s)\, ds - \frac{\lambda}{2\pi}\int_0^{2\pi} H(z(s))\, ds$$

as in §9.2. Now we apply the Lyapunov-Schmidt reduction to $f: \mathbb{R} \times Z \to Y$ at $(\lambda_0, 0)$ with $\lambda_0 = 2\pi/\beta$. This leads to the bifurcation equation

$$f(\lambda, v + h(v)) = f_\lambda(v + h(v)) = 0$$

which is defined in a neighborhood of $(\lambda_0, 0)$ in $\mathbb{R} \times V$ where $V = \ker Df_{\lambda_0}(0)$ as in §9.2. Here $h(v) \in V^\perp \subset Z$ is obtained via the implicit function theorem as usual. Moreover, it can be shown that the zeroes of the reduced equation are precisely the critical points of the functional

$$V \ni v \mapsto F_\lambda(v + h(v)) = A(v + h(v)) - \lambda B(v + h(v)) \in \mathbb{R}.$$

For details we refer the reader to [FaR2].

Using the isomorphism between V and E which associates to $v \in V$ the initial value $u = v(0)$ we have to find critical points of a functional $\Phi_\lambda = \Phi_\lambda(u)$ defined in a neighborhood of $(\lambda_0, 0)$ in $\mathbb{R} \times E$. Now all constructions are equivariant with respect to the action of Γ on Z, Y, V and E. And Γ-orbits of critical points of Φ_λ correspond to G-orbits of closed trajectories of (HS) with period λ. We want to apply the bifurcation theory of §7.5 to the negative gradient flow φ_λ on E associated to Φ_λ. Since there does not exist a reasonable version of the length for Γ we work instead with the length ℓ defined for the maximal torus $T\Gamma$ of Γ. To apply the results of §7.5 we need to know the exit-length $\ell^u(0, \varphi_\lambda)$ for λ near λ_0. According to formula (8.47) of [FaR2] we have

$$\Phi_\lambda(u) = \frac{1}{2}(\lambda_0 - \lambda)H''(0)u \cdot u + o(\|u\|^2)$$

for (λ, u) near $(\lambda_0, 0)$. This implies $\Phi''_\lambda(0) = (\lambda_0 - \lambda)H''(0)$ and therefore

$$\ell^u(0, \varphi_\lambda) = \begin{cases} \frac{1}{2} \cdot \dim E^+ & \text{for } \lambda < \lambda_0, \\ \frac{1}{2} \cdot \dim E^- & \text{for } \lambda > \lambda_0. \end{cases}$$

We should remark that Φ_λ is only C^1 in general because f_λ and, hence, h is only C^1. Still $\Phi''_\lambda(0)$ exists. This is a nice property of the Lyapunov-Schmidt reduction. If $0 \in E$ is not an isolated stationary solution of φ_{λ_0} then part (i) of Theorem 9.5 is true. Now suppose that 0 is an isolated stationary solution of φ_{λ_0}. Then the theorems 7.10 and 5.1 imply the existence of two integers $\tilde{d}_l, \tilde{d}_r \geq 0$ with

$$\tilde{d}_l + \tilde{d}_r \geq \frac{1}{2}|\dim E^+ - \dim E^-| = \frac{1}{2}\text{sgn}(H''(0)|E)$$

so that for each $\lambda < \lambda_0$ (respectively $\lambda > \lambda_0$) sufficiently close to λ_0 at least one of the following holds:

— There exist infinitely many critical $T\Gamma$-orbits of Φ_λ with different critical values.
— There exists an isolated invariant set S_λ of $E - 0$ with $\ell\big(\mathcal{C}(S_\lambda)\big) \geq \widetilde{d}_l$ (respectively $\ell\big(\mathcal{C}(S_\lambda)\big) \geq \widetilde{d}_r$).

If $\lambda \to \lambda_0$ then the critical points and the invariant sets converge to $0 \in E$. In the first case we must have infinitely many Γ-orbits of critical values of Φ_λ. And in the second case Theorem 6.1 tells us that S_λ must contain at least $d_l := \big\langle \widetilde{d}_l/\mu \big\rangle$ respectively $d_r := \big\langle \widetilde{d}_r/\mu \big\rangle$ Γ-orbits of stationary solutions of φ_λ. Here $\langle \alpha \rangle$ denotes the least integer greater than or equal to α.   □

9.6 Remark:

If $H''(0)|E$ is positive (or negative) definite then $\big|\mathrm{sgn}\big(H''(0)|E\big)\big| = \dim E$. In that case both theorems 9.2 and 9.5 are applicable. If part (i) of Theorem 9.1 does not hold then one expects that the number $d_l + d_r$ of G-orbits of closed trajectories parametrized by the period is the same as the number of such orbits on energy surfaces near the origin. Comparing 9.2 and 9.5 we see that we obtain the same lower bound if the connected component of G is a torus. In general, our results in the fixed energy case are better than those in the fixed period case. This is due to the fact that in the first case we can work with the $\widetilde{\mathcal{A}}_{SE}$-category for $T\Gamma$ or with the length ℓ for $T\Gamma$; see Remark 9.3. In the fixed period case, on the other hand, we do not have this choice. There we have to use ℓ because we do not have a category (or genus) version of the bifurcation theory of §7.5. From this point of view this somewhat unsatisfactory state of affairs is of a technical nature. Still it seems worthwhile to find a version which contains both theorems 9.2 and 9.5.

9.4 Symmetric periodic solutions

We want to refine the results of the preceding sections by looking at the symmetry of the periodic solutions. Consider a Hamiltonian function $H: \mathbb{R}^{2n} \to \mathbb{R}$ satisfying $(H3)$, that is H is invariant under a generalized symplectic action of a compact Lie group G on $\mathbb{R}^{2n}$. If $z: \mathbb{R} \to \mathbb{R}^{2n}$ is a 2π-periodic function solving

$$(HS_\lambda) \qquad\qquad \dot{z}(s) = \frac{\lambda}{2\pi} J \nabla H\big(z(s)\big)$$

then we call the isotropy group

$$\Gamma_z = \big\{ (\theta, g) \in \Gamma = S^1 \rtimes G : gz\big(\sigma(g)(s + \theta)\big) = z(s) \quad \text{for } s \in \mathbb{R} \big\}$$

the symmetry of z. The projection $\pi: \Gamma \to G$ maps Γ_z to the (spatial) symmetry $G_\tau = \pi(\Gamma_z)$ of the trajectory $\tau = z(\mathbb{R})$: $G_\tau = \{g \in G : g\tau = \tau\}$. Different symmetries Γ_z may lead to the same spatial symmetry G_τ.

Now suppose that H also satisfies the hypothesis $(H1)$ from §9.2. As usual let $E \subset \mathbb{R}^{2n}$ be the generalized eigenspace of $JH''(0)$ associated to the eigenvalues

$\pm ik\beta$, $k \in \mathbb{N}$. Recall that Γ acts on E. Fix an isotropy group Δ of $E - 0$. This is precisely the symmetry of a 2π-periodic solution of the linearized system

$$(LHS_{\lambda_0}) \qquad\qquad \dot{v}(t) = \frac{\lambda_0}{2\pi} JH''(0)v(t)$$

where $\lambda_0 = 2\pi/\beta$. Indeed, each non-zero element of the fixed point set E^Δ gives rise to a periodic solution of (LHS_{λ_0}) whose symmetry contains Δ. Now we ask how many of these periodic solutions remain if we perturb this linear system and look instead at (HS_λ) for λ near λ_0. Of course, we only want to count the number of Γ-orbits of periodic solutions of (LHS_λ) with symmetry at least (Δ), where (Δ) is the conjugacy class of Δ in Γ.

9.7 Theorem:

Suppose H satisfies (H1), (H2) and (H3) as in §9.2. Fix an isotropy group $\Delta \subset \Gamma$ of $E - 0$ and assume that the normalizer of Δ in Γ contains the subgroup $S^1 \cong S^1 \times \{1\}$ of $\Gamma = S^1 \rtimes G$. Then for sufficiently small $\varepsilon > 0$ there exist at least

$$\langle (\dim E^\Delta)/2(1 + \dim W\Delta - \operatorname{rk} W\Delta)\rangle$$

Γ-orbits of periodic solutions of (HS) on $H^{-1}(\varepsilon)$ with periods near $2\pi/\beta$ and symmetry at least (Δ).

The assumption $S^1 \subset N\Delta$ holds for all Δ if the action of G on $\mathbb{R}^{2n}$ is symplectic because in that case $\Gamma = S^1 \times G$ and therefore S^1 is in the center of Γ. This is not true in general. Examples can be found in 9.10 and 9.11 below. Theorem 9.2 corresponds to the special case where Δ is the trivial subgroup of Γ, so that $W\Delta = \Gamma$. Hypothesis $(H2)$ can be relaxed in 9.7. Instead of assuming $H''(0)$ to be positive definite on E it suffices to assume that $H''(0)|E^\Delta$ is positive definite.

Proof of Theorem 9.7:

As in the proof of Theorem 9.2 the problem is reduced to that of finding Γ-orbits of critical points of a Γ-invariant functional $\alpha: Z_\varepsilon \to \mathbb{R}$ where Z_ε is radially diffeomorphic to the unit sphere SE of E. We then obtain the solutions with the required symmetry if we look for critical points of the restriction $\alpha^\Delta: Z_\varepsilon^\Delta \to \mathbb{R}$. Since S^1 acts without fixed points on $E - 0$ we see that S^1 cannot be contained in Δ. Now the assumption $S^1 \subset N\Delta$ implies that the maximal torus of $W\Delta$ acts on $Z_\varepsilon^\Delta \subset E^\Delta$ without fixed points. Using Theorem 2.19 and Corollary 2.21 we can argue as in the proof of 9.2. $\qquad\square$

9.8 Remark:

The assumption $S^1 \subset N\Delta$ in Theorem 9.7 is true if and only if the projection $H := \pi(\Delta) \subset G$ acts symplectically on $\mathbb{R}^{2n}$, that is if $\sigma|H \equiv +1$. We treat the case $S^1 \not\subset N\Delta$ later. The case of symplectic actions has also been studied by Montaldi, Roberts and Stewart in [MoRS]. Their Theorem 1.1 guarantees the existence of $\frac{1}{2} \dim E^\Delta$ S^1-orbits of periodic solutions of (HS) on $H^{-1}(\varepsilon)$ with symmetry at least Δ. It is not clear how many Γ-orbits of periodic solutions with symmetry Δ exist. So the existence result of [MoRS] is weaker than Theorem 9.7. If $\dim W\Delta = 1$ then we obtain $\frac{1}{2} \dim E^\Delta$ Γ-orbits of periodic solutions on $H^{-1}(\varepsilon)$ with symmetry at least (Δ). And if $\dim W\Delta > 1$ then each such Γ-orbit contains infinitely many S^1-orbits. Therefore Theorem 1.1 of [MoRS] is only useful for those Δ which have a one-dimensional Weyl group in Γ.

The main results of [MoRS] are concerned with the (spectral) stability of the periodic solutions. In particular, Montaldi et al. investigate the effect of the symmetry on the Floquet operators. They also study various examples of symplectic group actions, especially for $G = D_n, SO(2), O(2)$ and $SU(2)$ acting on low dimensional spaces $E \subset \mathbb{R}^{2n}$. And they have some results for $G = O(3)$ acting on $E \cong V_l \oplus V_l \cong V_l \otimes \mathbb{C}$ where V_l is the space of spherical harmonics of order l. The S^1-action on E is given by scalar multiplication on the $\mathbb{C}$-component which means that no multiple of $i\beta$ is an eigenvalue of $JH''(0)$. Unfortunately, the isotropy groups $\Delta \subset \Gamma = S^1 \rtimes O(3)$ of $E - 0$ and the dimensions of the fixed point spaces have not been determined so far — except for those Δ with $\dim E^\Delta = 2$; see [GoS], §14, or [MoRS], §8. Therefore we do not attempt to study the case $G = O(3)$ in greater detail. Instead we look at some examples of generalized symplectic actions.

We consider first the action from 9.1b) where $G = \{1, I_1\} \cong \mathbb{Z}/2$ acts on $\mathbb{R}^{2n}$ via $I_1(p, q) = (-p, q)$. All classical Hamiltonian functions $H(p, q) = \frac{1}{2}\|p\|^2 + V(q)$ are invariant under this action. A special type of periodic solutions of (HS_λ) are the so-called *brake orbits*. These satisfy by definition $z(s) = I_1 z(-s)$; so the p-component is odd and the q-component is even about 0. Since z is 2π-periodic the p-component is odd and the q-component is even about π, too. For the corresponding λ-periodic solution $x(t) = z(2\pi t/\lambda)$ of (HS) the p-component is odd and the q-component is even about 0 and $\lambda/2$. Thus x is determined by its behavior between $t = 0$ and $t = \lambda/2$.

9.9 Theorem:

Suppose $H: \mathbb{R}^{2n} \to \mathbb{R}$ satisfies (H1), (H2) and (H3) with $G = \{1, I_1\}$ as above. Then the following holds.

a) *If no even multiple of $i\beta$ is an eigenvalue of $JH''(0)$ then (HS) has at least $\frac{1}{2} \dim E$ geometrically different brake orbits on $H^{-1}(\varepsilon)$ with periods near $2\pi/\beta$ for $\varepsilon > 0$ sufficiently small.*

b) *In general there exists at least one brake orbit of (HS) on $H^{-1}(\varepsilon)$ for $\varepsilon > 0$ small. If (HS) has precisely k brake orbits on $H^{-1}(\varepsilon)$ and $k < \frac{1}{2} \dim E$ then it has in addition at least $\dim E - 2k$ geometrically different periodic solutions on $H^{-1}(\varepsilon)$ which are not brake orbits. They appear in pairs $x(t), I_1 x(-t)$.*

A comment on this result is in order. The usual Weinstein-Moser theorem yields $m := \frac{1}{2} \dim E$ closed trajectories of (HS) on $H^{-1}(\varepsilon)$. Theorem 9.9 provides the additional information that either the corresponding periodic solutions are brake orbits or there are more than $\frac{1}{2} \dim E$ closed trajectories on $H^{-1}(\varepsilon)$. More precisely, if the number of brake orbits is k then $k \geq 1$ and there exist at least $\max\{m, 2m-k\}$ closed trajectories on $H^{-1}(\varepsilon)$. If $E = \mathbb{R}^{2n}$ and no integer multiple of $i\beta$ is an eigenvalue of $JH''(0)$ then part a) of Theorem 9.9 is a consequence of a result of Szulkin; see [Sz2], Theorem 4.1. The existence of at least one brake orbit on $H^{-1}(\varepsilon)$ follows also from a result of Rabinowitz [Ra3]. One can generalize 9.9 somewhat by replacing I_1 by $I_1 \circ I$ where I is a symplectic involution on $\mathbb{R}^{2n}$; see [BC2], Example 6.2.

Proof of Theorem 9.9:

a) First observe that $\Gamma = S^1 \rtimes G$ is isomorphic to $O(2)$. We set

$$\Delta = \big\{(0,1),(0,I_1)\big\} \subset \Gamma$$

which is isomorphic to G. Clearly, $z \in E^\Delta$ iff z is a brake orbit. The normalizer of Δ in Γ is given by

$$N\Delta = \big\{(0,1),(0,I_1),(\pi,1),(\pi,I_1)\big\} = \Delta \cup (\pi,1) \cdot \Delta.$$

Thus $z \in E^{N\Delta}$ implies $z(s+\pi) = \big((\pi,1) \cdot z\big)(s) = z(s)$. Since no even multiple of $i\beta$ is an eigenvalue of $JH''(0)$ the elements of $E - 0$ cannot have period π, so $E^{N\Delta} = 0$. This means that the Weyl group $W\Delta = N\Delta/\Delta \cong \mathbb{Z}/2$ acts on $Z_\varepsilon^\Delta \subset E^\Delta$ without fixed points. The $W\Delta$-equivariant map $Z_\varepsilon^\Delta \to \mathbb{R}$ from the proof of Theorem 9.7 must have at least $\dim E^\Delta$ critical $W\Delta$-orbits. Now 9.9a) follows from $\dim E^\Delta = \frac{1}{2} \dim E$.

b) The existence of at least one brake orbit is a consequence of part a) applied to the eigenvalue $2li\beta$ of $JH''(0)$ where $l \geq 1$ is maximal. To obtain the remaining solutions let $\Delta \subset \Gamma$ be the trivial group. Then $W\Delta = N\Delta = \Gamma$ and we can apply Theorem 9.7. We obtain at least $\frac{1}{2} \dim E$ Γ-orbits of periodic solutions of (HS) on $H^{-1}(\varepsilon)$ with periods near $2\pi/\beta$. Now a Γ-orbit can correspond either to one brake orbit or to two periodic solutions $z(s)$ and $I_1 z(-s)$. $\qquad\square$

9.10 Example:

Here we sketch a similar result for the action from 9.1 c) where

$$G = \big\{1, I_1, I_2, I_1 I_2\big\} \cong \mathbb{Z}/2 \times \mathbb{Z}/2$$

acts on $\mathbb{R}^{2n}$ via $I_1(p,q) = (-p,q)$ and $I_2(p,q) = (p,-q)$. For Hamiltonian systems with this symmetry (for example $H(p,q) = \frac{1}{2}\|p\|^2 + V(q)$ and V is even) so-called normal mode solutions occur (see [vGr]). A *normal mode solution* $z = (p,q)$ of (HS_λ) is a 2π-periodic brake orbit which satisfies in addition $z(\pi - s) = I_2 z(s)$. The corresponding λ-periodic solution $x(t) = z(2\pi t/\lambda)$ of (HS) satisfies

$$x\big(\tfrac{\lambda}{2} - t\big) = I_2 x(t).$$

Consequently, a normal mode solution is completely determined by its behavior between time 0 and one quarter of the period. For this action the group $\Gamma = S^1 \rtimes G$ is isomorphic to $O(2) \times \mathbb{Z}/2$. Setting $\Delta = \big\{(0,1),(0,I_1),(\pi,I_2),(\pi,I_1 I_2)\big\} \subset \Gamma$ one checks easily that $z \in E$ is a normal mode solution iff $z \in E^\Delta$. The normalizer of Δ in Γ is $N\Delta = \Delta \cup (\pi, 1) \cdot \Delta$, hence $W\Delta = N\Delta/\Delta \cong \mathbb{Z}/2$ similar to the situation in 9.9. If no even multiple of $i\beta$ is an eigenvalue of $JH''(0)$ then (HS) has at least $\frac{1}{2} \dim E$ normal mode solutions on $H^{-1}(\varepsilon)$ with period $2\pi/\beta$ for sufficiently small $\varepsilon > 0$. Observe that

$$\Delta' = \{(0,1),(0,I_2),(\pi,I_1),(\pi,I_1 I_2)\}$$

is conjugate to Δ: $\Delta' = \left(\frac{\pi}{2}, I_1\right) \cdot \Delta \cdot \left(\frac{\pi}{2}, I_1\right)^{-1}$. So working with Δ' instead of Δ yields the same Γ-orbits of periodic solutions. In general there exists at least one normal mode solution of (HS) on $H^{-1}(\varepsilon)$. Theorem 9.7 applied to the trivial subgroup of Γ yields at least $\frac{1}{2} \dim E$ Γ-orbits of periodic solutions on $H^{-1}(\varepsilon)$. Such a Γ-orbit corresponds either to one normal mode solution or to two brake orbits which are not normal modes or to four geometrically different periodic solutions which are not brake orbits. Such a quadruple looks like $x(t), I_2 x(t), I_1 x(-t), -x(-t)$. We leave the details to the reader.

9.11 Remark:

Clearly $(\pi, 1) \in \Gamma$ is in the center of Γ no matter what G and $\sigma: G \to \{\pm 1\}$ look like. A straightforward computation shows that for every subgroup Δ of Γ either $S^1 \cap N\Delta = S^1$ or $S^1 \cap N\Delta = \{(0,1),(\pi,1)\} \cong \mathbb{Z}/2$; to see this one uses the fact that Δ is determined by a triple (H, θ, d) where H is a subgroup of G, $\theta: H \to S^1$ is a homomorphism and $d \in \mathbb{N}$:

$$\Delta = \{(\theta(h) + k/d, h): h \in H, \ k = 0, 1, \ldots, d-1\}.$$

In the case $S^1 \cap N\Delta = S^1$ we can apply Theorem 9.7. In the second case $S^1 \cap N\Delta \cong \mathbb{Z}/2$ we can use the methods from the last chapter provided no even multiple of $i\beta$ is an eigenvalue of $JH''(0)$. This means that we study the negative gradient flow of $\alpha: Z_\varepsilon \to \mathbb{R}$ with the help of the $\hat{\mathcal{A}}$-category for $\mathbb{Z}/2$ where $\mathcal{A} = \{\mathbb{Z}/2\}$. In addition we use the facts that this flow leaves the fixed point spaces Z_ε^Δ invariant for every $\Delta \subset \Gamma$ and that the critical orbits of α form $\mathbb{Z}/2$-manifolds of the form Γ/Δ. Of course this requires some knowledge of the isotropy groups Δ occuring in $E - 0$ and of the dimensions of the fixed point spaces E^Δ. Moreover, one should know the $\hat{\mathcal{A}}$-category of the orbit spaces Γ/Δ. For reasons of computation it may be simpler to work with the genus or the length for $\mathbb{Z}/2$ instead of the category.

It is also possible to generalize Theorem 9.5 by taking the symmetry of the periodic solutions into account. For simplicity we shall only formulate the analogue of 9.7 and leave those of 9.8 to 9.11 to the reader as well as the proof of the following result.

9.12 Theorem:
Suppose $H: \mathbb{R}^{2n} \to \mathbb{R}$ satisfies $(H1)$, $(H2')$ and $(H3)$ as in §9.3. Fix an isotropy group $\Delta \subset \Gamma$ of $E - 0$ and assume that the normalizer of Δ in Γ contains S^1. Then one of the following holds:

(i) *There exists a sequence z_i of periodic solutions of (HS) with period $2\pi/\beta$ and $z_i \to 0$ as $i \to \infty$.*

(ii) *There exist integers $d_l(\Delta), d_r(\Delta) \geq 0$ with*

$$d_l(\Delta) + d_r(\Delta) \geq \left| \operatorname{sgn}\left(H''(0)|E^\Delta\right)/2\mu(\Delta) \right|$$

and having the property: For each $\lambda < 2\pi/\beta$ (respectively $\lambda > 2\pi/\beta$) and λ sufficiently close to $2\pi/\beta$ there exist at least $d_l(\Delta)$ (respectively $d_r(\Delta)$) Γ-orbits of periodic solutions of (HS_λ) with symmetry at least (Δ). These periodic solutions converge towards the origin as $\lambda \to 2\pi/\beta$. Here

$$\mu(\Delta) = \max\{\ell_\Delta(W\Delta u): u \in E^\Delta - 0\}$$

where ℓ_Δ is the length for the maximal torus of $W\Delta$ (see 4.3).   □

Another way to obtain periodic solutions near an equilibrium is to apply Hopf bifurcation theorems as done by Alexander and Yorke [AlY], for instance. Equivariant versions can be found in the paper [GoS] by Golubitsky and Stewart and the book [Fi] by Fiedler. These results can be applied to more general equivariant vector fields than Hamiltonian systems. In addition they yield global branches of periodic solutions (see [Fi]). On the other hand they do not give the multiplicity statements of 9.2, 9.5, 9.7 etc.

References

[AlY]   Alexander, J.C., and Yorke, J.A., Global bifurcation of periodic orbits. *Amer. J. Math.* **100** (1978), 263-292.

[AmR]   Ambrosetti, A., and Rabinowitz, P.H., Dual variational methods in critical point theory. *J. Funct. Anal.* **14** (1973), 349-381.

[At]   Atiyah, M.F., *Lectures on K-Theory.* Benjamin, New York 1967.

[AtT]   Atiyah, M.F., and Tall, D.O., Group representations, λ-rings and the *J*-homomorphism. *Topology* **8** (1969), 253-298.

[Ba1]   Bartsch, T., Critical orbits of symmetric functionals. *Manuscr. Math.* **66** (1989), 129-152.

[Ba2]   Bartsch, T., On the genus of representation spheres. *Comment. Math. Helv.* **65** (1990), 85-95.

[Ba3]   Bartsch, T., On the existence of Borsuk-Ulam theorems. *Topology* **31** (1992), 533-543.

[Ba4]   Bartsch, T., A simple proof of the degree formula for $\mathbb{Z}/p$-equivariant maps. *Math. Z.* **212** (1993), 285-292.

[Ba5]   Bartsch, T., Infinitely many solutions of a symmetric Dirichlet problem. *Nonl. Anal., Theory, Meth. & Appl.* **20** (1993), 1205-1216.

[BC1]   Bartsch, T., and Clapp, M., Bifurcation theory for symmetric potential operators and the equivariant cup-length. *Math. Z.* **204** (1990), 341-356.

[BC2]   Bartsch, T., and Clapp, M., The compact category and multiple periodic solutions of Hamiltonian systems on symmetric starshaped energy surfaces. *Math. Ann.* **293** (1992), 523-542.

[BCP]   Bartsch, T., Clapp, M., and Puppe, D., A mountain pass theorem for actions of compact Lie groups. *J. reine angew. Math.* **419** (1991), 55-66.

[Be1]   Benci, V., On critical point theory for indefinite functionals in the presence of symmetries. *Trans. Amer. Math. Soc.* **274** (1982), 533-572.

[Be2]   Benci, V., A geometrical index for the group S^1 and some applications to the study of periodic solutions of ordinary differential equations. *Commun. Pure Appl. Math.* **34** (1981), 393-432.

[BeP]   Benci, V., and Pacella, F., Morse theory for symmetric functionals on the sphere and an application to a bifurcation problem. *Nonl. Anal., Theory, Meth. & Appl.* **9** (1985), 763-773.

[BeR]   Benci, V., and Rabinowitz, P.H., Critical point theorems for indefinite functionals. *Invent. Math.* **52** (1979), 241-273.

[BLMR]   Berestycki, H., Lasry, J.M., Mancini, G., and Ruf, B., Existence of multiple periodic orbits on starshaped Hamiltonian surfaces. *Comm. Pure Appl. Math.* **38** (1985), 253-289.

[Bi]   Birkhoff, G.D., Dynamical systems with two degrees of freedom. *Transactions Amer. Math. Soc.* **18** (1917), 199-300.

[Böh]   Böhme, R., Die Lösung der Verzweigungsgleichungen für nichtlineare Eigenwertprobleme. *Math. Z.* **127** (1972), 105-126.

[Bor]   Borel, A., *Seminar on Transformation Groups.* Princeton University Press, Princeton 1960.

[Bre]   Bredon, G.E., *Introduction to Compact Transformation Groups.* Academic Press, New York 1972.

[BröD]   Bröcker, T., and tom Dieck, T., *Representations of Compact Lie Groups.* Springer, New York 1985.

[Bro]   Browder, F.E., Nonlinear eigenvalue problems and group invariance. In: *Functional Analysis and Related Fields*, F.E. Browder (ed.), Proc. Chicago 1968, Springer 1970, 1-58.

[Bu]   Busse, F.H., Pattern of convection in spherical shells. *J. Fluid Mech.* **72** (1975), 65-85.

[Ca]   Carlsson, G., Equivariant stable homotopy and Segal's Burnside ring conjecture. *Ann. Math.* **120** (1984), 189-224.

[CLM]   Chossat, P., Lauterbach, R., and Melbourne, I., Steady-state bifurcation with $O(3)$-symmetry. *Arch. Rat. Mech. and Anal.* **113** (1990), 313-376.

[ChL]   Chow, S.-N., and Lauterbach, R., A bifurcation theorem for critical points of variational problems. *Nonl. Anal., Theory, Meth. & Appl.* **12** (1988), 51-61.

[ClP1]   Clapp, M., and Puppe, D., Invariants of Lusternik-Schnirelmann type and the topology of critical sets. *Transactions Amer. Math. Soc.* **298** (1986), 603-620.

[ClP2]   Clapp, M., and Puppe, D., Critical point theory with symmetries. *J. reine angew. Math.* **418** (1991), 1-29.

[Cof]   Coffman, C.V., A minimum-maximum principle for a class of nonlinear integral equations. *J. Analyse Math.* **22** (1969), 391-419.

[Con]   Conley, C., *Isolated invariant sets and the Morse index.* CBMS, Regional Confer. Ser. in Math. 38, Amer. Math. Soc., Providence, R.I., 1978.

[ConZ]   Conley, C., and Zehnder, E., A Morse type index theory for flows and periodic solutions to Hamiltonian systems. *Comm. Pure Appl. Math.* **37** (1984), 207-253.

[ConnF]   Conner, P.E., and Floyd, E.E., Fixed point free involutions and equivariant maps I. *Bull. Amer. Math. Soc.* **66** (1960), 416-441; II. *Transactions Amer. Math. Soc.* **105** (1962), 222-228.

[CorDM]   Corvellec, J.-N., Degiovanni, M., and Marzocchi, M., Deformation properties for continuous functionals and critical point theory. *Top. Meth. Nonlinear Anal.* **1** (1993), 151-171.

[Di]   tom Dieck, T., *Transformation Groups.* de Gruyter, Berlin 1987.

[Do]   Dold, A., *Lectures on Algebraic Topology.* Grundlehren der mathematischen Wissenschaften 200, Springer, Berlin 1980.

[Dr]   Dress, A., A characterization of solvable groups. *Math. Z.* **110** (1969), 213-217.

[Ei]   Eilenberg, S., On a theorem of P.A. Smith concerning fixed points for
            periodic transformations. *Duke Math. J.* **6** (1940), 428-437.

[EiS]   Eilenberg, S., and Steenrod, N.E., *Foundations of Algebraic Topology.*
            Princeton University Press, Princeton 1952.

[EkL]   Ekeland, I., and Lasry, J.M., On the number of periodic trajectories for
            a Hamiltonian flow on a convex energy surface. *Ann. Math.* **112** (1980),
            283-319.

[Fa1]   Fadell, E., The relationship between Ljusternik-Schnirelmann category
            and the concept of genus. *Pacific J. Math.* **89** (1980), 33-42.

[Fa2]   Fadell, E., The equivariant Lusternik-Schnirelmann method for invari-
            ant functionals and relative cohomological index theories. In: *Méth.
            topologiques en analyse non linéaires*, A. Granas (ed.), Sémin. Math. Sup.
            No. **95** Montréal 1985, 41-70.

[Fa3]   Fadell, E., Cohomological methods in non-free G-spaces with applications
            to general Borsuk-Ulam theorems and critical point theorems for invariant
            functionals. In: *Nonlinear Functional Analysis and its Applications*, S.P.
            Singh (ed.), Proc. Maratea 1985, NATO ASI Ser. **C 173** Reidel, Dordrecht
            1986, 1-45.

[FaH1]   Fadell, E., and Husseini, S., Relative cohomological index theories. *Ad-
            vances in Math.* **64** (1987), 1-31.

[FaH2]   Fadell, E., and Husseini, S., An ideal valued cohomological index theory
            with applications to Borsuk-Ulam and Bourgain-Yang theorems. *Ergod.
            Th. & Dynam. Sys.* 8* (1988), 73-85.

[FaHR]   Fadell, E., Husseini, S., and Rabinowitz, P.H., Borsuk-Ulam theorems for
            S^1-actions and applications. *Transactions Amer. Math. Soc.* **274** (1982),
            345-359.

[FaR1]   Fadell, E., and Rabinowitz, P.H., Bifurcation for odd potential operators
            and an alternative topological index. *J. Funct. Anal.* **26** (1977), 48-67.

[FaR2]   Fadell, E., and Rabinowitz, P.H., Generalized cohomological index theo-
            ries for Lie group actions with an application to bifurcation questions for
            Hamiltonian systems. *Inv. Math.* **45** (1978), 139-174.

[Fe]   Feit, W., *Characters of Finite Groups.* W.A. Benjamin, New York 1967.

[Fi]   Fiedler, B., *Global Bifurcation of Periodic Solutions with Symmetry.* Lec-
            ture Notes in Mathematics 1309, Springer, Berlin 1988.

[FiM]   Fiedler, B., and Mischaikow, K., Dynamics of bifurcations for variational
            problems with O(3)-equivariance: a Conley index approach. *Arch. Rat.
            Mech. and Anal.* **119** (1992), 145-196.

[FiP]   Fiedorowicz, Z., and Priddy, S., *Homology of Classical Groups Over Fi-
            nite Fields and Their Associated Infinite Loop Spaces.* Lecture Notes in
            Mathematics 674, Springer, Berlin 1978.

[Field]   Field, M.J., Equivariant bifurcation theory and symmetry breaking. *Dyn.
            Diff. Eqns.* **1** (1989), 369-421.

[Fl]   Floer, A., A refinement of the Conley index and an application to the stability of hyperbolic invariant sets. *Ergod. Th. and Dynam. Sys.* **7** (1988), 93-103.

[FlZ]   Floer, A., and Zehnder, E., The equivariant Conley index and bifurcations of periodic solutions of Hamiltonian systems. *Ergod. Th. and Dynam. Syst.* **8** (1988), 87-97.

[FLRW]   Fournier, G., Lupo, D., Ramos, M., and Willem, M., Limit relative category and critical point theory. *Dynamics Reported* (to appear).

[FoW]   Fournier, G., and Willem, M., Multiple solution of the forced double pendulum equation. *Ann. Inst. H. Poincaré, Analyse non linéaire* **6** suppl. (1989), 259-281.

[Fr]   Franzosa, R.D., The connection matrix theory for Morse decompositions. *Transactions Amer. Math. Soc.* **311** (1989), 561-592.

[Gi]   Girardi, M., Multiple orbits for Hamiltonian systems on starshaped surfaces with symmetries. *Ann. Inst. H. Poincaré, Nouv. Sér., Sect. B1, vol.* **1** (1984), 285-294.

[GoS]   Golubitsky, M., and Stewart, I.N., Hopf bifurcation in the presence of symmetry. *Arch. Rat. Mech. and Anal.* **87** (1985), 107-165.

[Gor]   Gorenstein, D., *Finite Groups.* Harper & Row, New York-Evanston-London 1968.

[vGr]   van Groesen, E.W.C., Existence of multiple normal mode trajectories on convex energy surfaces of even, classical Hamiltonian systems. *J. Diff. Equ.* **57** (1985), 70-89.

[Gro]   Grothendieck, A., Sur quelques points d'algèbre homologique. *Tohoku Math. J.* **9** (1957), 119-221.

[Henn]   Henn, H.-W., Cohomological p-nilpotence criteria for compact Lie groups. In: *Colloque sur la théorie de l'homotopie* Luminy 1988. *Astérisque* **191** (1990), 211-220.

[Henry]   Henry, D., *Geometric Theory of Semilinear Parabolic Equations.* Lecture Notes in Mathematics 840, Springer, Berlin 1981.

[HPS]   Hirsch, M.W., Pugh, C.C., and Shub, M., *Invariant Manifolds.* Springer, Berlin 1977.

[Ho]   Hochschild, G.P., *The Structure of Lie Groups.* Holden-Day, San Francisco 1965.

[Hs]   Hsiang, W.Y., *Cohomology Theory of Topological Transformation Groups.* Ergebnisse der Mathematik und ihrer Grenzgebiete, Bd. 85, Springer, Berlin 1975.

[Hu]   Huppert, B., *Endliche Gruppen I.* Grundlehren der math. Wiss., Bd. 143, Springer, Berlin 1967.

[HuB]   Huppert, B., and Blackburn, N., *Finite Groups III.* Grundlehren der math. Wiss., Bd. 243, Springer, Berlin 1982.

[IG]   Ihrig, E., and Golubitsky, M., Pattern selection with O(3) symmetry. *Physica* **13D** (1984), 1-33.

[Il]   Illman, S., The equivariant triangulation theorem for actions of compact
          Lie groups. *Math. Ann.* **262** (1983), 487-501.

[IM]   Izydorek, M., and Marzantowicz, W., Equivariant maps between represen-
          tation spheres. *Heidelberg, Forschungsschwerpunkt Geometrie, Heft Nr.* **70**
          (1990).

[Ka]   Kato, T., *Perturbation Theory for Linear Operators.* Grundlehren der
          mathematischen Wissenschaften 132, Springer, Berlin 1976.

[KnS]   Knightly, G.H., and Sather, D., Buckled states of a spherical shell under
          uniform external pressure. *Arch. Rat. Mech. and Anal.* **72** (1980), 315-380.

[Ko1]   Komiya, K., Equivariant critical point theory and ideal-valued cohomo-
          logical index.   Preprint.

[Ko2]   Komiya, K., Equivariant critical point theory and set-valued genus.
          Preprint.

[Kr1]   Krasnoselski, M.A., On special coverings of a finite-dimensional sphere.
          (in Russian) Dokl. Akad. Nauk SSSR **103** (1955), 966-969.

[Kr2]   Krasnoselski, M.A., *Topological Methods in the Theory of Nonlinear In-
          tegral Equations.* Pergamon Press, Oxford 1964.

[KMP]   Krasnoselski, M.A., Muhamediev. E.M., and Pokrovski, A.V., Bifurcation
          values of parameters in variational problems. *Sov. Math. Dokl.* **22** (1980),
          682-686.

[Lai]   Laitinen, E., On the Burnside ring and stable cohomotopy of a finite group.
          Math. Scand. **44** (1979), 37-72.

[LaiM]   Laitinen, E., and Morimoto, M., Finite groups with smooth one fixed point
          actions on spheres.   Preprint.

[Lau]   Lauterbach, R., *Problems with Spherical Symmetry: Studies on Bifurca-
          tion and Dynamics for O(3)-Equivariant Equations.* Habilitationsschrift,
          Augsburg 1988.

[LeW]   Lee, C.-N., and Wasserman, A.G., *On the groups JO(G).* Mem. Amer.
          Math. Soc. **159**, Providence, R.I., 1975.

[LMS]   Lewis, L.G., May, J.P., and Steinberger, M., *Equivariant Stable Homotopy
          Theory.* Lecture Notes in Mathematics 1213, Springer, Berlin 1986.

[Lu]   Lusternik, L., Topologische Grundlagen der allgemeinen Eigenwerttheorie.
          Monatsh. Math. Phys. **37** (1930), 125-130.

[LuS1]   Lusternik, L., and Schnirelmann, L., Sur le problème de trois géodésique
          fermées sur les surfaces de genre 0. *C.R.Acad. Sci. Paris* **189** (1929), 269-
          271.

[LuS2]   Lusternik, L., and Schnirelmann, L., *Méthodes topologiques dans les
          problèmes variationels.* Hermann, Paris 1934.

[Ly]   Lyapunov, A.M., Problème général de la stabilité du mouvement. *Ann.
          Fac. Sci. Toulouse* **2** (1907), 203-474.

[Ma]   Marino, A., La biforcazione nel caso variazionale. *Proc. Conf. Sem. Mat.
          Univ. Bari* **132** (1977).

[Mar]   Marzantowicz, W., An almost classification of compact Lie groups with Borsuk-Ulam properties. *Pacific J. Math.* **144** (1990), 299-311.

[Mat]   Matumoto, T., On G-CW complexes and a theorem of J.H.C. Whitehead. *J. Fac. Sci. Tokyo* **18** (1971/72), 363-374.

[MaW]   Mawhin, J., and Willem, M., *Critical Point Theory and Hamiltonian Systems.* Springer, New York 1989.

[Me]   Meyer, D., Zur Existenz von $\mathbb{Z}/p$-äquivarianten Abbildungen zwischen Linsenräumen und Sphären. Diplomarbeit Universität Heidelberg, Heidelberg 1993.

[Mi]   Michalek, R., A $\mathbb{Z}^p$-Borsuk-Ulam theorem and index theory with a multiplicity result in partial differential equations. *Nonl. Anal., Theory, Meth. & Appl.* **13** (1989), 957-986.

[MiT]   Michalek, R., and Tarantello, G., Subharmonic solutions with prescribed minimal period for nonautonomous Hamiltonian systems. *J. Diff. Eq.* **72** (1988), 28-55.

[MilS]   Milnor, J.W., and Stasheff, J.D., *Characteristic Classes.* Annals of Mathematics Studies, Number **76**, Princeton University Press, Princeton 1974.

[MoRS]   Montaldi, J.A., Roberts, R.M., and Stewart, I.N., Periodic solutions near equilibria of symmetric Hamiltonian systems. *Phil. Trans. R. Soc. London, A325, No.* **1584** (1988), 237-293.

[Mor1]   Morse, M., The critical points of functions and the calculus of variations in the large. *Bull. Amer. Math. Soc.* **35** (1929), 38-54.

[Mor2]   Morse, M., *The Calculus of Variations in the Large.* Colloq. Publ. Vol. **18**, Amer. Math. Soc., Providence, R.I., 1934.

[Mos]   Moser, J., Periodic orbits near an equilibrium and a theorem by A. Weinstein. *Comm. Pure Appl. Math.* **29** (1976), 727-747.

[Mun]   Munkholm, H.J., On the Borsuk-Ulam theorem for $\mathbb{Z}_{p^a}$ actions on S^{2n-1} and maps $S^{2n-1} \to \mathbb{R}$. *Osaka J. Math.* **7** (1970), 451-456.

[Mur]   Murayama, M., On G-ANRs and their G-homotopy type. *Osaka J. Math.* **20** (1983), 479-512.

[P1]   Palais, R., Homotopy theory of infinite dimensional manifolds. *Topology* **5** (1966), 1-16.

[P2]   Palais, R., Lusternik-Schnirelman theory on Banach manifolds. *Topology* **5** (1966), 115-132.

[Pazy]   Pazy, A., *Semigroups of Linear Operators and Applications to Partial Differential Equations.* Springer, New York 1983.

[Po]   Pontriagin, L., *Topological Groups.* Princeton University Press, Princeton 1946.

[Ra1]   Rabinowitz, P.H., Some global results for nonlinear eigenvalue problems. *J. Func. Anal.* **7** (1971), 487-513.

[Ra2]   Rabinowitz, P.H., *Minimax Methods in Critical Point Theory With Applications to Differential Equations.* CBMS, Regional Confer. Ser. in Math. **65**, Amer. Math. Soc., Providence, R.I., 1986.

[Ra3]   Rabinowitz, P.H., On the existence of periodic solutions for a class of symmetric Hamiltonian systems. *Nonl. Anal., Theory, Meth. & Appl.* **11** (1987), 599-611.

[Ram]   Ramsay, J.R., *Extensions of Ljusternik-Schnirelmann Category Theory to Relative, Equivariant and Isovariant Theories.* Ph. D. Thesis, University of Wisconsin-Madison, 1982.

[Re1]   Reeken, M., Stability of critical points under small perturbations. Part I: Topological theory. *Manus. math.* **7** (1972), 387-411.

[Re2]   Reeken, M., Stability of critical points under small perturbations. Part II: Analytical theory. *Manus. math.* **8** (1973), 69-92.

[Ry]   Rybakowski, K.P., *The Homotopy Index and Partial Differential Equations.* Springer, Berlin 1987.

[Sa]   Salamon, D., Connected simple systems and the Conley index of isolated invariant sets. *Trans. Amer. Math. Soc.* **291** (1985), 1-41.

[Sc]   Schnirelmann, L., Über eine neue kombinatorische Invariante. *Monatsh. Math. Phys.* **37** (1930), 131-134.

[Se1]   Segal, G., Equivariant K-theory. *Publ. Math. IHES* **34** (1968), 129-151.

[Se2]   Segal, G., Equivariant stable homotopy theory. *Actes Congr. internat. Math. 1970,* **2** 59-63.

[Si1]   Singhof, W., On the Lusternik-Schnirelmann category of Lie groups. *Math. Z.* **145** (1975), 111-116.

[Si2]   Singhof, W., On the Lusternik-Schnirelmann category of Lie groups, II. *Math. Z.* **151** (1976), 143-148.

[Sp]   Spanier, E., *Algebraic Topology.* Mc-Graw Hill, New York 1966.

[Ste]   Steinlein, H., Borsuk's antipodal theorem and its generalizations and applications: a survey. In: *Méth. topologiques en analyse non linéaires,* A. Granas (ed.), Sémin. Math. Sup. No. **95** Montréal 1985, 166-235.

[Sto]   Stolz, S., The level of real projective spaces. *Comment. Math. Helv.* **64** (1989), 661-674.

[Str]   Struwe, M., *Variational Methods.* Springer, Berlin 1990.

[Su]   Suzuki, M., On a class of doubly transitive groups. *Ann. of Math.* **75** (1962), 104-145.

[Sv]   Švarc, A.S., The genus of fibre spaces. *Trudy Moskov Mat. Obšč.* **10** (1961), 217-272, and **11** (1962), 99-126 (in Russian); English translation in *Transl. Amer. Math. Soc., II. Ser.,* **55** (1966), 49-140.

[Sw]   Switzer, R.M., *Algebraic Topology — Homotopy and Homology.* Springer, Berlin 1975.

[Sz1]   Szulkin, A., Ljusternik-Schnirelmann theory on C^1-manifolds. *Ann. Inst. H. Poincaré, Analyse non linéaire* **5** (1988), 119-139.

[Sz2]   Szulkin, A., An index theory and existence of multiple brake orbits for star-shaped Hamiltonian systems. *Math. Ann.* **283** (1989), 241-255.

[Sz3]   Szulkin, A., A relative category and applications to critical point theory for strongly indefinite functionals. *J. Nonl. Anal.* **15** (1990), 725-739.

[Ta]   Tarantello, G., Subharmonic solutions for Hamiltonian systems via a $\mathbb{Z}_p$ pseudoindex theory.   Preprint.

[To]   Tornehave, J., Equivariant maps of spheres with conjugate orthogonal actions. In: *Algebraic Topology, Proc. Conf., London Ont. 1981, Canadian Math. Soc. Conf. Proc.* **2**, part 2, (1982), 275-301.

[VI]   Vanderbauwhede, A., and Iooss, G., Center manifold theory in infinite dimensions. In: *Dynamics Reported. New Series* (C.K.R.T. Jones, U. Kirchgraber, H.O. Walther eds.), Vol. **1**, Springer, Berlin 1992, 125-163.

[Vick] Vick, J., An application of K-theory to equivariant maps. *Bull. Amer. Math. Soc.* **75** (1969), 1017-1019.

[Wa1]   Waner, S., Equivariant homotopy theory and Milnor's theorem. *Transactions Amer. Math. Soc.* **258** (1980), 351-368.

[Wa2]   Waner, S., A note on the existence of G-maps between spheres. *Proc. Amer. Math. Soc.* **99** (1987), 179-181.

[Wang] Wang, Z.-Q., A $\mathbb{Z}_p$-Borsuk-Ulam theorem. *Chinese Science Bulletin* **34**, No. 14, (1989), 1153-1157.

[Was]   Wasserman, A.G., Isovariant maps and the Borsuk-Ulam theorem. *Topology and its applications* **38** (1991), 155-161.

[We]   Weinstein, A., Normal modes for nonlinear Hamiltonian systems. *Invent. Math.* **20** (1973), 47-57.

[Wh]   Whitehead, G.W., *Elements of Homotopy Theory.* Springer, New York 1978.

[Wo]   Wolf, J.A., *Spaces of constant curvature.* McGraw-Hill, New York 1967.

[Y1]   Yang, C.T., On the theorems of Borsuk-Ulam, Kakutani-Yamabe-Yujòbô and Dyson I. *Ann. Math.* **60** (1954), 262-282.

[Y2]   Yang, C.T., On the theorems of Borsuk-Ulam, Kakutani-Yamabe-Yujòbô and Dyson II. *Ann. Math.* **62** (1955), 271-283.

[Y3]   Yang, C.T., Continuous functions from spheres to Euclidean spaces. *Ann. Math.* **62** (1955), 284-292.

[Z]   Zeidler, E., *Nonlinear Functional Analysis and its Applications III.* Springer, New York 1988.

Subject Index

Printing: Weihert-Druck GmbH, Darmstadt
Binding: Buchbinderei Schäffer, Grünstadt

Vol. 1463: R. Roberts, I. Stewart (Eds.), Singularity Theory and its Applications. Warwick 1989, Part II. VIII, 322 pages. 1991.

Vol. 1464: D. L. Burkholder, E. Pardoux, A. Sznitman, Ecole d'Eté de Probabilités de Saint- Flour XIX-1989. Editor: P. L. Hennequin. VI, 256 pages. 1991.

Vol. 1465: G. David, Wavelets and Singular Integrals on Curves and Surfaces. X, 107 pages. 1991.

Vol. 1466: W. Banaszczyk, Additive Subgroups of Topological Vector Spaces. VII, 178 pages. 1991.

Vol. 1467: W. M. Schmidt, Diophantine Approximations and Diophantine Equations. VIII, 217 pages. 1991.

Vol. 1468: J. Noguchi, T. Ohsawa (Eds.), Prospects in Complex Geometry. Proceedings, 1989. VII, 421 pages. 1991.

Vol. 1469: J. Lindenstrauss, V. D. Milman (Eds.), Geometric Aspects of Functional Analysis. Seminar 1989-90. XI, 191 pages. 1991.

Vol. 1470: E. Odell, H. Rosenthal (Eds.), Functional Analysis. Proceedings, 1987-89. VII, 199 pages. 1991.

Vol. 1471: A. A. Panchishkin, Non-Archimedean L-Functions of Siegel and Hilbert Modular Forms. VII, 157 pages. 1991.

Vol. 1472: T. T. Nielsen, Bose Algebras: The Complex and Real Wave Representations. V, 132 pages. 1991.

Vol. 1473: Y. Hino, S. Murakami, T. Naito, Functional Differential Equations with Infinite Delay. X, 317 pages. 1991.

Vol. 1474: S. Jackowski, B. Oliver, K. Pawałowski (Eds.), Algebraic Topology, Poznań 1989. Proceedings. VIII, 397 pages. 1991.

Vol. 1475: S. Busenberg, M. Martelli (Eds.), Delay Differential Equations and Dynamical Systems. Proceedings, 1990. VIII, 249 pages. 1991.

Vol. 1476: M. Bekkali, Topics in Set Theory. VII, 120 pages. 1991.

Vol. 1477: R. Jajte, Strong Limit Theorems in Noncommutative L_2-Spaces. X, 113 pages. 1991.

Vol. 1478: M.-P. Malliavin (Ed.), Topics in Invariant Theory. Seminar 1989-1990. VI, 272 pages. 1991.

Vol. 1479: S. Bloch, I. Dolgachev, W. Fulton (Eds.), Algebraic Geometry. Proceedings, 1989. VII, 300 pages. 1991.

Vol. 1480: F. Dumortier, R. Roussarie, J. Sotomayor, H. Żoładek, Bifurcations of Planar Vector Fields: Nilpotent Singularities and Abelian Integrals. VIII, 226 pages. 1991.

Vol. 1481: D. Ferus, U. Pinkall, U. Simon, B. Wegner (Eds.), Global Differential Geometry and Global Analysis. Proceedings, 1991. VIII, 283 pages. 1991.

Vol. 1482: J. Chabrowski, The Dirichlet Problem with L^2-Boundary Data for Elliptic Linear Equations. VI, 173 pages. 1991.

Vol. 1483: E. Reithmeier, Periodic Solutions of Nonlinear Dynamical Systems. VI, 171 pages. 1991.

Vol. 1484: H. Delfs, Homology of Locally Semialgebraic Spaces. IX, 136 pages. 1991.

Vol. 1485: J. Azéma, P. A. Meyer, M. Yor (Eds.), Séminaire de Probabilités XXV. VIII, 440 pages. 1991.

Vol. 1486: L. Arnold, H. Crauel, J.-P. Eckmann (Eds.), Lyapunov Exponents. Proceedings, 1990. VIII, 365 pages. 1991.

Vol. 1487: E. Freitag, Singular Modular Forms and Theta Relations. VI, 172 pages. 1991.

Vol. 1488: A. Carboni, M. C. Pedicchio, G. Rosolini (Eds.), Category Theory. Proceedings, 1990. VII, 494 pages. 1991.

Vol. 1489: A. Mielke, Hamiltonian and Lagrangian Flows on Center Manifolds. X, 140 pages. 1991.

Vol. 1490: K. Metsch, Linear Spaces with Few Lines. XIII, 196 pages. 1991.

Vol. 1491: E. Lluis-Puebla, J.-L. Loday, H. Gillet, C. Soulé, V. Snaith, Higher Algebraic K-Theory: an overview. IX, 164 pages. 1992.

Vol. 1492: K. R. Wicks, Fractals and Hyperspaces. VIII, 168 pages. 1991.

Vol. 1493: E. Benoît (Ed.), Dynamic Bifurcations. Proceedings, Luminy 1990. VII, 219 pages. 1991.

Vol. 1494: M.-T. Cheng, X.-W. Zhou, D.-G. Deng (Eds.), Harmonic Analysis. Proceedings, 1988. IX, 226 pages. 1991.

Vol. 1495: J. M. Bony, G. Grubb, L. Hörmander, H. Komatsu, J. Sjöstrand, Microlocal Analysis and Applications. Montecatini Terme, 1989. Editors: L. Cattabriga, L. Rodino. VII, 349 pages. 1991.

Vol. 1496: C. Foias, B. Francis, J. W. Helton, H. Kwakernaak, J. B. Pearson, H_∞-Control Theory. Como, 1990. Editors: E. Mosca, L. Pandolfi. VII, 336 pages. 1991.

Vol. 1497: G. T. Herman, A. K. Louis, F. Natterer (Eds.), Mathematical Methods in Tomography. Proceedings 1990. X, 268 pages. 1991.

Vol. 1498: R. Lang, Spectral Theory of Random Schrödinger Operators. X, 125 pages. 1991.

Vol. 1499: K. Taira, Boundary Value Problems and Markov Processes. IX, 132 pages. 1991.

Vol. 1500: J.-P. Serre, Lie Algebras and Lie Groups. VII, 168 pages. 1992.

Vol. 1501: A. De Masi, E. Presutti, Mathematical Methods for Hydrodynamic Limits. IX, 196 pages. 1991.

Vol. 1502: C. Simpson, Asymptotic Behavior of Monodromy. V, 139 pages. 1991.

Vol. 1503: S. Shokranian, The Selberg-Arthur Trace Formula (Lectures by J. Arthur). VII, 97 pages. 1991.

Vol. 1504: J. Cheeger, M. Gromov, C. Okonek, P. Pansu, Geometric Topology: Recent Developments. Editors: P. de Bartolomeis, F. Tricerri. VII, 197 pages. 1991.

Vol. 1505: K. Kajitani, T. Nishitani, The Hyperbolic Cauchy Problem. VII, 168 pages. 1991.

Vol. 1506: A. Buium, Differential Algebraic Groups of Finite Dimension. XV, 145 pages. 1992.

Vol. 1507: K. Hulek, T. Peternell, M. Schneider, F.-O. Schreyer (Eds.), Complex Algebraic Varieties. Proceedings, 1990. VII, 179 pages. 1992.

Vol. 1508: M. Vuorinen (Ed.), Quasiconformal Space Mappings. A Collection of Surveys 1960-1990. IX, 148 pages. 1992.

Vol. 1509: J. Aguadé, M. Castellet, F. R. Cohen (Eds.), Algebraic Topology - Homotopy and Group Cohomology. Proceedings, 1990. X, 330 pages. 1992.

Vol. 1510: P. P. Kulish (Ed.), Quantum Groups. Proceedings, 1990. XII, 398 pages. 1992.

Vol. 1511: B. S. Yadav, D. Singh (Eds.), Functional Analysis and Operator Theory. Proceedings, 1990. VIII, 223 pages. 1992.

Vol. 1512: L. M. Adleman, M.-D. A. Huang, Primality Testing and Abelian Varieties Over Finite Fields. VII, 142 pages. 1992.

Vol. 1513: L. S. Block, W. A. Coppel, Dynamics in One Dimension. VIII, 249 pages. 1992.

Vol. 1514: U. Krengel, K. Richter, V. Warstat (Eds.), Ergodic Theory and Related Topics III, Proceedings, 1990. VIII, 236 pages. 1992.

Vol. 1515: E. Ballico, F. Catanese, C. Ciliberto (Eds.), Classification of Irregular Varieties. Proceedings, 1990. VII, 149 pages. 1992.

Vol. 1516: R. A. Lorentz, Multivariate Birkhoff Interpolation. IX, 192 pages. 1992.

Vol. 1517: K. Keimel, W. Roth, Ordered Cones and Approximation. VI, 134 pages. 1992.

Vol. 1518: H. Stichtenoth, M. A. Tsfasman (Eds.), Coding Theory and Algebraic Geometry. Proceedings, 1991. VIII, 223 pages. 1992.

Vol. 1519: M. W. Short, The Primitive Soluble Permutation Groups of Degree less than 256. IX, 145 pages. 1992.

Vol. 1520: Yu. G. Borisovich, Yu. E. Gliklikh (Eds.), Global Analysis – Studies and Applications V. VII, 284 pages. 1992.

Vol. 1521: S. Busenberg, B. Forte, H. K. Kuiken, Mathematical Modelling of Industrial Process. Bari, 1990. Editors: V. Capasso, A. Fasano. VII, 162 pages. 1992.

Vol. 1522: J.-M. Delort, F. B. I. Transformation. VII, 101 pages. 1992.

Vol. 1523: W. Xue, Rings with Morita Duality. X, 168 pages. 1992.

Vol. 1524: M. Coste, L. Mahé, M.-F. Roy (Eds.), Real Algebraic Geometry. Proceedings, 1991. VIII, 418 pages. 1992.

Vol. 1525: C. Casacuberta, M. Castellet (Eds.), Mathematical Research Today and Tomorrow. VII, 112 pages. 1992.

Vol. 1526: J. Azéma, P. A. Meyer, M. Yor (Eds.), Séminaire de Probabilités XXVI. X, 633 pages. 1992.

Vol. 1527: M. I. Freidlin, J.-F. Le Gall, Ecole d'Eté de Probabilités de Saint-Flour XX – 1990. Editor: P. L. Hennequin. VIII, 244 pages. 1992.

Vol. 1528: G. Isac, Complementarity Problems. VI, 297 pages. 1992.

Vol. 1529: J. van Neerven, The Adjoint of a Semigroup of Linear Operators. X, 195 pages. 1992.

Vol. 1530: J. G. Heywood, K. Masuda, R. Rautmann, S. A. Solonnikov (Eds.), The Navier-Stokes Equations II – Theory and Numerical Methods. IX, 322 pages. 1992.

Vol. 1531: M. Stoer, Design of Survivable Networks. IV, 206 pages. 1992.

Vol. 1532: J. F. Colombeau, Multiplication of Distributions. X, 184 pages. 1992.

Vol. 1533: P. Jipsen, H. Rose, Varieties of Lattices. X, 162 pages. 1992.

Vol. 1534: C. Greither, Cyclic Galois Extensions of Commutative Rings. X, 145 pages. 1992.

Vol. 1535: A. B. Evans, Orthomorphism Graphs of Groups. VIII, 114 pages. 1992.

Vol. 1536: M. K. Kwong, A. Zettl, Norm Inequalities for Derivatives and Differences. VII, 150 pages. 1992.

Vol. 1537: P. Fitzpatrick, M. Martelli, J. Mawhin, R. Nussbaum, Topological Methods for Ordinary Differential Equations. Montecatini Terme, 1991. Editors: M. Furi, P. Zecca. VII, 218 pages. 1993.

Vol. 1538: P.-A. Meyer, Quantum Probability for Probabilists. X, 287 pages. 1993.

Vol. 1539: M. Coornaert, A. Papadopoulos, Symbolic Dynamics and Hyperbolic Groups. VIII, 138 pages. 1993.

Vol. 1540: H. Komatsu (Ed.), Functional Analysis and Related Topics, 1991. Proceedings. XXI, 413 pages. 1993.

Vol. 1541: D. A. Dawson, B. Maisonneuve, J. Spencer, Ecole d´ Eté de Probabilités de Saint-Flour XXI - 1991. Editor: P. L. Hennequin. VIII, 356 pages. 1993.

Vol. 1542: J.Fröhlich, Th.Kerler, Quantum Groups, Quantum Categories and Quantum Field Theory. VII, 431 pages. 1993.

Vol. 1543: A. L. Dontchev, T. Zolezzi, Well-Posed Optimization Problems. XII, 421 pages. 1993.

Vol. 1544: M.Schürmann, White Noise on Bialgebras. VII, 146 pages. 1993.

Vol. 1545: J. Morgan, K. O'Grady, Differential Topology of Complex Surfaces. VIII, 224 pages. 1993.

Vol. 1546: V. V. Kalashnikov, V. M. Zolotarev (Eds.), Stability Problems for Stochastic Models. Proceedings, 1991. VIII, 229 pages. 1993.

Vol. 1547: P. Harmand, D. Werner, W. Werner, M-ideals in Banach Spaces and Banach Algebras. VIII, 387 pages. 1993.

Vol. 1548: T. Urabe, Dynkin Graphs and Quadrilateral Singularities. VI, 233 pages. 1993.

Vol. 1549: G. Vainikko, Multidimensional Weakly Singular Integral Equations. XI, 159 pages. 1993.

Vol. 1550: A. A. Gonchar, E. B. Saff (Eds.), Methods of Approximation Theory in Complex Analysis and Mathematical Physics IV, 222 pages, 1993.

Vol. 1551: L. Arkeryd, P. L. Lions, P.A. Markowich, S.R. S. Varadhan. Nonequilibrium Problems in Many-Particle Systems. Montecatini, 1992. Editors: C. Cercignani, M. Pulvirenti. VII, 158 pages 1993.

Vol. 1552: J. Hilgert, K.-H. Neeb, Lie Semigroups and their Applications. XII, 315 pages. 1993.

Vol. 1553: J.-L- Colliot-Thélène, J. Kato, P. Vojta. Arithmetic Algebraic Geometry. Editor: E. Ballico. VII, 223 pages. 1993.

Vol. 1554: A. K. Lenstra, H. W. Lenstra, Jr. (Eds.), The Development of the Number Field Sieve. VIII, 131 pages. 1993.

Vol. 1555: O. Liess, Conical Refraction and Higher Microlocalization. X, 389 pages. 1993.

Vol. 1556: S. B. Kuksin, Nearly Integrable Infinite-Dimensional Hamiltonian Systems. XXVII, 101 pages. 1993.

Vol. 1557: J. Azéma, P. A. Meyer, M. Yor (Eds.), Séminaire de Probabilités XXVII. VI, 327 pages. 1993.

Vol. 1558: T. J. Bridges, J. E. Furter, Singularity Theory and Equivariant Symplectic Maps.

Vol. 1560: T. Bartsch, Topological Methods for Variational Problems with Symmetries. X, 152 pages. 1993.